T0319565

Managing Technology and Product Development Programmes

Managing Technology and Product Development Programmes

A Framework for Success

Peter Flinn
Ashbourne, UK

This edition first published 2019

Registered Offices
John Wiley & Sons, Inc., 111 River Street, Hoboken, NJ 07030, USA
John Wiley & Sons Ltd, The Atrium, Southern Gate, Chichester, West Sussex, PO19 8SQ, UK

Editorial Office
The Atrium, Southern Gate, Chichester, West Sussex, PO19 8SQ, UK

For details of our global editorial offices, customer services, and more information about Wiley products visit us at www.wiley.com.

Wiley also publishes its books in a variety of electronic formats and by print-on-demand. Some content that appears in standard print versions of this book may not be available in other formats.

Library of Congress Cataloging-in-Publication Data

Names: Flinn, Peter, 1948- author.
Title: Managing technology and product development programmes : a framework for success / Mr. Peter Flinn, Ashbourne, UK.
Description: Hoboken, NJ : John Wiley & Sons, Inc., 2019. | Includes bibliographical references and index. |
Identifiers: LCCN 2018044854 (print) | LCCN 2018046236 (ebook) | ISBN 9781119517269 (Adobe PDF) | ISBN 9781119517252 (ePub) | ISBN 9781119517245 (hardcover)
Subjects: LCSH: New products.
Classification: LCC TS170 (ebook) | LCC TS170 .F55 2019 (print) | DDC 658.5/75–dc23
LC record available at https://lccn.loc.gov/2018044854

Cover Design: Wiley
Cover Image: © Alexander Supertramp / Shutterstock

Set in 10/12pt WarnockPro by SPi Global, Chennai, India

Printed and bound by CPI Group (UK) Ltd, Croydon, CR0 4YY

10 9 8 7 6 5 4 3 2 1

About the Author

Peter Flinn is a British engineer who originally studied mechanical sciences at the University of Cambridge. He also has an MBA and studied international management at Harvard University. He is a Fellow of the Institution of Mechanical Engineers.

During his career, he worked in the aerospace, commercial vehicle, rail, and process industries holding chief engineer, head of engineering, and managing director positions within international organisations. In recent years, he has led the creation of the Manufacturing Technology Centre in Coventry, and the Aerospace Technology Institute in Cranfield, both in the United Kingdom.

Throughout his nearly 50-year career, he has taken a keen, practical interest in the subject of this book – how to develop technology and products. He has direct experience of technology research work through all phases of development to manufacturing management. The content is based on this experience and, in particular, on what does or does not produce successful results. He hopes that the content of the book will prove useful to engineers, technologists, and investors in these fields.

Contents

1

Introduction

1.1 Why Write This Book?

Most aspiring engineers would like to see their name attached to a product, such as a car or plane, or a structure, such as a building or bridge – they want to make their mark. In the early stages of their careers, their contribution might just be a minor element of the whole; later, they would hope to take the lead, possibly even emulating Isambard Kingdom Brunel, Sir Frank Whittle, or Steve Jobs. But how does a product get from the glimmer of an idea to the finished item?

This book is concerned with the way that new research and technology ideas are converted into products that can be manufactured and sold to satisfied customers. Its emphasis is on engineered products but the principles can be applied more widely. It might be thought that this subject would already have extensive coverage, given that engineering has been taught in European countries as a degree-level subject for over 200 years.

École Polytechnique in Paris, for example, was set up in 1794 specifically to address, amongst other things, the dearth of qualified engineers at that time (Ref. 1.1). The University of Glasgow was the first in the United Kingdom to set up a school of engineering with the appointment of Lewis Gordon (1815–1876) as Regius Professor of Civil Engineering and Mechanics at the University. He was in the post from 1840 until 1855, when he resigned to pursue his successful business interests – thus providing an early demonstration of the economic value of engineering (Ref. 1.2).

Engineering has been taught in Cambridge, where the author studied, since 1875, albeit sometimes using a different title such as 'mechanical sciences'. But for some 100 years beforehand, 'real and useful knowledge' was taught as an extension of mathematics and covering such topics as steam engines and mechanisms (Ref. 1.3).

The content of such academically demanding courses has been biased towards the science of engineering, with an emphasis on mathematical analysis, although more practical and applied skills have also been covered

Managing Technology and Product Development Programmes: A Framework for Success,
First Edition. Peter Flinn.
© 2019 John Wiley & Sons Ltd. Published 2019 by John Wiley & Sons Ltd.

through laboratory work, as well as through design and build projects. Those interested in the business aspects of engineering can then go on to acquire formal qualifications in this area, such as an MBA. This is a well-trodden path for those wishing to pursue a managerial career in the engineering business.

However, there has been relatively little coverage of the processes by which engineering products are created and developed. These processes use the design, analysis, and other skills that are taught academically. However, the means by which a technology is turned into a design through to its being launched as a, hopefully, reliable product to a discerning public is something that a new engineer has to work out for him or herself – a process that can often take a full decade of puzzlement if the environment is complicated.

The management thinker Peter Drucker drew attention to this topic in a 1985 essay in which he stated: 'We know how to train people to do technology such as engineering or chemistry. But we do not know how to endow managers with *technological literacy*, that is, with an understanding of technology and its dynamics…Yet technological literacy is increasingly a major requirement for managers ….'

The purpose of this book, then, is to help fill this gap, as illustrated in Figure 1.1, by providing a framework that can be used to describe how new technologies, and then products, are created.

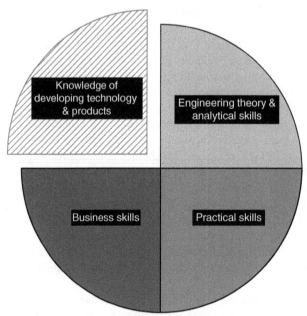

Figure 1.1 Context of the material in this book.

The word *framework* is used advisedly. It would be wrong to suggest that there is a universal, reliable, and prescriptive set of rules for developing technology; it is a far more hit-and-miss process, as will be described later. However, there are principles and approaches that can make the process more efficient and that engineers typically learn by trial and error.

1.2   Importance of the Product Development Process

There is widespread interest in this topic, and in innovation more widely, in business schools and similar organisations (Ref. 1.4). Their work has shown a clear link between the effectiveness of companies' product development processes and the overall performance of those companies. A number of large-scale academic studies support these conclusions, and this has led to the considerable interest in the topic of innovation, which, as an academic topic, has been widely researched but essentially from a business, financial, and marketing perspective. Quite rightly, organisations are continually exhorted to 'innovate or die!'

However, these studies rarely link back into the engineering processes that drive the actual creation of the products and that represent the practical realisation of innovation.

Hence, there is a strong business incentive for companies to become more proficient in renewing their products or services. The recognition that successful products achieve then provides an incentive to well-rounded engineers making their vital contribution to this topic.

In terms of written material to support the development of proficiency, there is a significant body of literature on single topics within the product renewal process. For example, there are weighty books on risk management, design for manufacture, engineering analysis, and project management, to name but a few. However, there is little tying these topics together. The purpose of this book is to fill that void.

1.3   Perspective of This Book

Given this background and these early references to engineering topics, this book examines the processes for developing new products and technologies from an engineering perspective. It is based on the author's 45–50 years of practical experience in engineering industries, including aerospace, automotive, rail, and some process industries. Whilst the emphasis is based on practical experience, and hence what works, there is considerable academic underpinning to the approaches suggested, and this is referenced whenever possible. The book should not, however, be regarded as 'research-based'; it is 'experience based', if such a category exists, derived from participation rather than observation.

A strong thread permeating the book is the linkage between those engineering processes and the wider performance of the business. The point is frequently made that technology is only worth developing if it can be put to practical use. It must be deployed in the form of products or services bought by paying customers at a price which is economically sustainable for it to be of any useful interest. The book will help engineers to understand how their contribution fits into the wider context of the business.

1.4 Intended Readership

This book addresses the topics above and is aimed at those who are still trying to understand the processes for turning science into technology and then into products. The readership could include:

- Engineers in their final stage of university education, perhaps undertaking final year, capstone projects, or MSc programmes
- Technologists or engineers in the early stages of their careers, particularly those working in industry on technology and product development
- Technology researchers who would like to understand more about the means by which their research work could eventually lead to commercial products
- Business school researchers who are working in the field of innovation
- Commercial managers, finance managers, and business people whose work involves managing, funding, or approving technology development but who do not necessarily have a direct involvement or direct experience
- Investors who might be asked to fund technology-based ventures
- Seasoned engineers and engineering managers, who will recognise most of the material in this book but who might find its content brings further structure to their thoughts

The language of the book is, of necessity, somewhat technical, as it is impossible to describe engineering processes without using some technical language. However, it is intended to be straightforward and easily understood – there is no arcane language, and there are no formulae!

1.5 Science, Technology, Innovation, Engineering, and Product Development

Some definitions are appropriate at this early stage. The terms *science, technology, innovation, engineering,* and *product development* are sometimes used interchangeably, with overlaps between the areas that they attempt to define. For the purposes of this book, the terms are defined as follows:

- *Science.* The systematic study of the structure and behaviour of the physical and natural world through observation and experiment.
- *Research.* The development, for its own sake, of new ideas, knowledge, and science without any firm application in mind.
- *Technology.* The practical application of research and science to develop new solutions that could subsequently be taken into commercial application through a product or service.
- *Product development.* The development of a specific, commercial product for use in the marketplace, probably using a combination of existing and new technology.

Engineering is, then, used as an umbrella term, described in the Oxford English Dictionary as 'the branch of science and technology concerned with the design, building, and use of engines, machines, and structures', or, alternatively 'the action of working artfully to bring something about'. The first definition is very logical and straightforward; the second, using the term *artfully*, suggests a hint of manipulation, but it is true that engineering is as much about engineering a result, by whatever means are available, as conducting a form of science.

Another frequently used word is *innovation*. The word is understood as meaning: 'to make changes in something established, especially by introducing new methods, ideas, or products'. It has a wider context and a slightly more modern ring to it than engineering. It does include turning ideas and technology into saleable products or services, but innovation can also include changes to business models, commercial arrangements, and services. Innovation is used quite widely in business school and government policy circles, where there is a recognition that companies must continually innovate to stay alive. *Innovation* is not used quite as frequently in this book but is, nonetheless, at the heart of what has been written.

Perhaps the most perceptive comment on this whole area comes from Arthur M. Wellington, an American civil engineer who wrote the 1887 book *The Economic Theory of the Location of Railways*. The saying that '*an engineer can do for a dollar what any fool can do for two*' is attributed to him.

1.6 The Changing Nature of Engineering

As suggested above, the development of technology and new products is the route by which new knowledge is converted to practical use. Arguably, the fundamentals of this process have remained constant since the first products were created many centuries ago.

However, the process is not insulated from the wider world and has itself been influenced quite significantly by new technology. Any review of the approaches and methods for developing technology and products must take these trends into account. In principle, they do not affect the fundamental nature of development processes. However, they do make them quicker, more productive, and more rigorous. Particular trends include:

- The replacement of conventional engineering information, such as drawings and textual information, by digital forms extending to product lifecycle management systems, which capture all product-related data
- The ability to share digitally common, and completely up-to-date, information across all business areas and functions, as well as with suppliers and customers
- The extension of analytical methods into highly representative mathematical models, reducing the need for physical testing
- The adoption of cleverer methods of physical testing using means that are more representative of real life and with more thorough and quicker data analysis

Allied to this, the nature of the manufacturing industry has also seen fundamental change, influenced also by technology and broader economic or political factors. Key trends in this respect have been and continue to be:

- Closer coupling of engineering to manufacturing, with the engineering focussed towards strategic manufacturing and broader business aims
- Whole-life considerations being given more prominence in all phases of the engineering process
- More emphasis on new technologies and ideas coming from outside the firm
- Stronger links to, and more cooperation with, suppliers, who are given greater responsibility and on whom greater dependence is placed
- More geographically distributed supply chains with more layers and greater specialisation
- Greater customer expectation of a tailor-made product – mass customisation
- Closer interaction between the supplying company and the products in the field, including direct data feed from operating products back to the manufacturer – the Internet of Things
- The effects of some products being sold as a service operated by the product supplier, rather than being sold to the end customer for self-operation – sometimes known as 'servitisation' – creating new models of business

The broader business processes by which firms develop new products have also received considerable attention through topics such as:

- Product success and failure factors
- Product innovation as a competitive weapon
- The marketing/new product development interface, teamwork, and integration of new product development activities

- Technology portfolio development and open innovation
- *Trans*-national new product development

These factors collectively add up to an environment of significant, and continuing, change to which engineering processes must respond.

1.7 The Fourth Industrial Revolution

These points also need to be seen in the context of what is being described as a fourth Industrial Revolution, revolving around data, embedded computing, digital methods, and artificial intelligence, as summarised in Figure 1.2. The term *Industry 4.0* is another frequently used title for this development.

Engineering, product development, and service development activities are at the heart of this fourth Industrial Revolution, both as an originator of new data and in analysing data from products in the field. Klaus Schwab's book [1.5] provides some interesting ideas about how the future might play out in this area over the coming decades.

1.8 Scope of This Book

The emphasis of this book is the process by which new products and technology are made to work, sometimes described as *validation*. This use of this word, however, can lead to a somewhat narrow understanding of the topic in the sense

Revolution	First	Second	Third	Fourth
Approximate Timescale	1760–1840	1880–1910	1965–1995	2000–
Features	Coal Steam engine Railways Mechanisation and factories Iron production	Electricity Steel Mass production Large corporations	Semiconductors Computing & software Internet Automation	Digital manufacturing Virtual modelling Embedded computing Artificial intelligence & autonomy

Figure 1.2 Four Industrial Revolutions.

that it relates more to a checking procedure after everything else has been finished. For example, one definition of validation (with some very minor wording changes) from ISO9000 is: 'an independent procedure that is used for checking that a product, service, or system meets requirements and specifications and that it fulfils its intended purpose'.

The thrust of this book, then, is more towards ensuring at every step along the way that a successful outcome will be achieved, from original conception to validation as the final confirmation – recognising, of course, that success is not guaranteed and that not every idea is viable. However, it considers how to ensure that a good design reaches its full potential, including its potential from a customer's perspective and its potential from the business perspective of the manufacturer.

1.9   Structure of This Book

After this introduction, the book begins with a description and analysis of the process of engineering – the steps by which new technology makes its way from the laboratory to the commercial marketplace. A particular emphasis is placed on the fact that such a process does exist but that it is not, unlike manufacturing processes, highly repetitive or linear in character. In particular, it does involve a strong element of discovery and learning.

This is followed by a discussion of technology maturity, describing the characteristics and status of technology at its different stages of development. It makes the point that, rather like the human life cycle, new technology has to mature through a process that cannot be shortcut without causing problems. This is not say, however, that the process cannot be accelerated or made more efficient if properly understood, but the right foundations do have to be created.

Chapter 4 is concerned with aligning technology and product development with wider manufacturing, commercial, and business considerations. The key, and important, point here is that technology on its own has limited value, beyond satisfying general curiosity. Its value comes from the creation of new products and services. This is where the importance of meeting customer needs comes into play and of doing this in a way that creates good business for the supplying organisation through efficient manufacturing and service operations.

The book then moves on to how technology and product development should be planned and organised. Despite the earlier comments about the relatively unstructured nature of the technology development process, there are ways in which it can be managed that give a much better chance of a satisfactory result being achieved.

This leads into Chapter 6, describing the development of new concepts – the early stages of the process, which are more fluid and less-easily defined than the later stages.

Chapter 7 examines the subject of risk. Although not always recognised in this way, the topic of risk effectively brings into play the lessons of the past. Risk, in the sense of avoiding the mistakes of the past, is integral to development processes and it needs to be actively managed. Poor management of risks leads to underdevelopment of products and a higher probability of problems and failures – failure of the product to function properly, failure to meet customer requirements, and failure to meet business targets. Identifying and overcoming risks is arguably the most important aspect of all development activities.

This is followed by coverage of development and validation: how those risks, once identified, can be analysed, overcome, and proven to be properly overcome so they don't cause further problems. The emphasis is on three types of validation activity: engineering calculation, modelling and simulation, and physical testing.

Validation precedes or runs in parallel with the main phase of engineering delivery, described in Chapter 9. *Delivery* here refers to all the product information required to manufacture, assemble, test, commission, and support the finished product – a phase of work that is much more structured and predictable than earlier phases.

Chapter 10 describes the important topic of how these programmes of work can be funded: an individual's own resources, company-generated cash flow, or externals investors, for example.

The next two main chapters concern human aspects of technology and product development. Chapter 11 covers the organisation of the people who undertake the real work – for example, how to run an engineering team, and how to work with other organisations, which could include suppliers, customers, research partners, shareholders, and advisers. It also describes aspects of leadership, recruitment, and personal development. The overriding thought is that the processes described in earlier chapters are run and managed by people. Hence, the results are only as good as the organisation itself and the way it handles the discovery, learning, and decision-making processes referred to earlier.

Chapter 12 moves into the area of critical thinking and decision-making – somewhat philosophical points, perhaps, but an important and very human aspect of engineering activities. Recent work in the social sciences, and in areas such as economics, has shown that human decision-making, which happens every day in engineering work, is subject to all sorts of biases and human foibles. There are no magic cures to this tendency, but it is useful to be aware of the likely pitfalls so they can, if at all possible, be avoided. There is also some discussion of methods for structuring and solving problems and for a more creative approach to problem solving.

Chapter 13 is concerned about implementing the processes described above in early-stage companies or improving the processes of existing organisations. 'Change Management' has emerged as an indispensable business topic in recent decades, reflecting the need for companies to adapt themselves to changing circumstances, or go out of business. Adaption of engineering development processes is as important as any other process and is not easy to undertake, given that its activities are intangible and involve wide groups of people. This chapter addresses some of these issues.

The concluding section then summarises the key messages of the book and offers some suggestions about how the future might play out for material covered in the book.

1.10   Reading Sequence

As already implied, the book is intended to be read in the chapter sequence described. However, each chapter is relatively self-contained and each describes material that, in most cases, is already the subject of weighty, single-topic books. The value of this book, however, is in tying these topics together in a hopefully logical framework, beginning with an understanding of the engineering process, which is probably the one major topic in this book which is not widely described elsewhere.

Then, the last section of the book is longer than might normally be the case. It could be read as a mini-book about technology and product development, covering a dozen or so pages.

References

The first three references provide some further historical background to the development of engineering as an academic topic in leading European universities.

1.1 Ecole Polytechnique — https://www.polytechnique.edu/en/key-dates
1.2 University of Glasgow — http://www.universitystory.gla.ac.uk/chair-and-lectureship/?id=711
1.3 Cambridge Engineering, The First 150 Years, Haroon Ahmed, 2017

This short essay looks at the development of product innovation as an academic research discipline by examining the number, and topic matter, of academic papers over approximately 30 years from 1984 onwards.

1.4 Anthony di Benedetto, C. (2013). The emergence of the product innovation discipline. In: *PDMA Handbook of New Product Development*, 3e. John Wiley.

Finally, this short book lays out its author's view of how the so-called Fourth Industrial Revolution might play out.

1.5 The Fourth Industrial Revolution – Klaus Schwab, World Economic Forum, 2016

2

Engineering as a Process

2.1 Background

The word *process* is used widely in the world of engineering, industry, and business. It generally refers to a sequence of activities that produce a result. The dictionary defines it as 'a series of actions or steps taken in order to achieve a particular end'. Similarly, ISO9001 talks of a 'set of interrelated or interacting activities that use inputs to deliver an intended result'.

Examples of a process might include a manufacturing process to produce a component, an administrative process to produce an invoice, or an 'HR' process to recruit someone. The steps can be defined and the process can then be mapped and measured, which can, in turn, lead to improvement in the performance of that process. Above all, such processes are repetitive and happen with a relatively high frequency. Hence, the results of process improvement come through quickly and their success can be judged in days or months.

Technology and product development is also a process, albeit a complex one. It is not, as some have argued, a journey without a map. However, it is not highly repetitive – the timescale from end-to-end can be years or even decades – and each programme is uniquely individual. Improvement is therefore more difficult to achieve; indeed, some engineers in certain industries may only see two or three complete cycles in their working life.

2.2 The Basic Components of the Process

This is not to say, however, that considering the process of technology and product development is a futile exercise. It can be broken down into its elements, and an approach can be developed for each phase. In fact, later in the development process, there are more elements that are repetitive and specific, lending themselves to classic analysis and improvement. The earlier steps remain, however, more elusive and justify their description as the 'fuzzy front end'.

Managing Technology and Product Development Programmes: A Framework for Success,
First Edition. Peter Flinn.
© 2019 John Wiley & Sons Ltd. Published 2019 by John Wiley & Sons Ltd.

For the purposes of this book, four phases of technology and product development have been identified:

1) Science
2) Technology research
3) Technology development
4) Product development

Science, in this context, is the process by which new knowledge is created for its own sake. For example, the science of semi-conduction, or its forebears, was discovered in the 1820s with the observation that the electrical resistance of some materials decreases with temperature. It was some time before any real use was made of this phenomenon. Karl Ferdinand Braun developed the crystal diode rectifier some 50 years later, providing the basis of the first cheap domestic radios. It was another 80 years before semiconduction started to hit the headlines with the development of the transistor in 1947 and all that then followed in the computing industry.

Technology research is then the process by which science is developed towards some useful application and is the start point for this book. The development of transistors in the 1940s might be considered part of this phase, although much of the work could also fall into the category of science.

Technology development takes this useful application and progresses it to the point where confidence in it is much higher, through greater understanding of the detail, and where a commercial enterprise then feels able to commit to developing and selling a product to the marketplace. Transistors went through this process initially in the late 1940s and early 1950s in a number of laboratories, mainly in the United States.

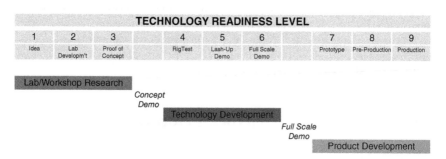

Figure 2.1 Main phases of development.

The final phase – product development – and the most expensive, is where a useful product, such as a transistor radio, is made and sold in volume – first achieved, albeit in a very crude way by today's standards, in the early- to mid-1950s.

The main phases of development are illustrated on Figure 2.1 whilst the details of the steps in maturing a new technology are described in much more detail in Chapter 3, which covers, in particular, the concept of 'technology readiness'.

2.3 Expenditure on Research and Development

The processes of science, technology development, and product development make up a significant proportion of economic activity. In the developed nations, somewhere between 1% and 4% of those nations' economic output is devoted to R&D, as illustrated in Figure 2.2, which covers some 40 countries of the Organisation for Economic Co-operation and Development (OECD).

These figures include expenditure by:

- Universities and other higher educational institutions
- Government
- Nonprofit private organisations
- Business (the largest contributor)

They also cover all forms of R&D, which are usually broken down by economists and funding organisations into three categories:

1) Basic research. For the advancement of knowledge.
2) Applied research. To acquire new knowledge, directed primarily towards a specific, practical aim or objective.
3) Experimental development. Directed towards specific new products and processes.

Globally, R&D spending was estimated in 2016 to total $1.95 trillion in purchasing power parity values for the more than 110 countries having significant R&D investments of over more than $100 million per year (Ref. 2.1).

This then plays into measures of innovation performance across nations, as ranked in Figure 2.3.

It can be seen, therefore, that technology development in its various forms is a significant activity in its own right. It is also argued that countries that spend a greater proportion of their GDP on these activities are also the most successful, economically, with 3% of GDP often being stated as a desirable target.

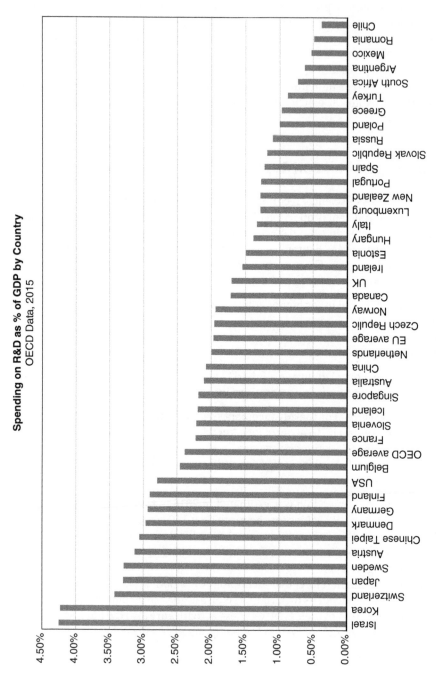

Figure 2.2 Expenditure on R&D as percentage of GDP for 41 OECD countries.

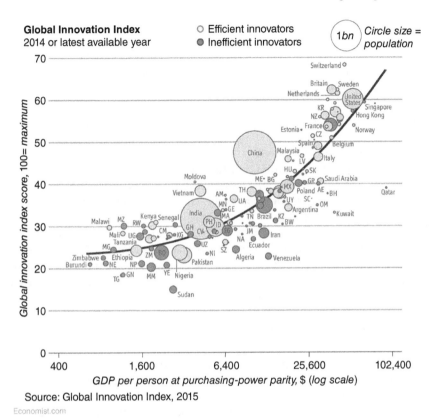

Figure 2.3 Global innovation index as published by *The Economist* – Ref. 2.2.

2.4 Economic Returns from R&D Work

There is a considerable body of academic work, probably unknown to most of the technology community, on the subject of linking R&D spending to economic success. Investment in R&D is like any other form of investment: resources are spent now in the expectation of future returns. However, unlike, say, the purchase of a machine tool, the returns can be some way off in the future, success is not guaranteed, and results may be difficult to attribute to one, specific piece of R&D. Nonetheless, work has been done to estimate the effect of R&D on sales, shareholder returns, market value, and margins (Ref. 2.3).

The results have a wide spread, as might be expected given the fact that any specific R&D project may be spectacularly successful or may be a complete failure. The 'private' returns to companies involved in manufactured products are estimated to be in the range 10–30% per annum. The term *private* here refers to the return experienced by the firm making the investment.

There is also the so-called *social return*, which refers to the wider benefit which industry more generally experiences from R&D work. Findings in one company tend, over time, to spill over into other companies in the same sector or to other sectors. Supplier companies, for example, will serve several customers and may supply across sectors. Individual research engineers acquire tacit knowledge from their experience that will guide them in future work as they move from job to job. (Patents can protect specific designs, but the engineers will take with them the experience of what does or doesn't work or what approaches might be successful).

Social returns to R&D are estimated to be in the range of 40–100% per annum in manufacturing industries. The substantial gap between private and social returns is one of the arguments advanced for government support for R&D through grants or tax credits.

Overall, then, the economic returns from technology and product development are high, which helps explain why the percentage of GDP spent on research is growing in most countries. At the same time, the variability of returns emphasises the need to select and manage R&D projects carefully, a subject covered in more detail in Chapter 6.

2.5   Science as the Precursor of Technology

As noted in Chapter 1, science creates the foundation for new technology and hence new products. A look back in time, when life was arguably simpler, apparently shows a direct link between the great scientific discoveries of the seventeenth to nineteenth centuries and their application by the great engineering characters of a slightly later era, as illustrated in Figure 2.4.

Further investigation suggests that the relationship between science and engineering is less straightforward than this timeline might suggest. For example, there seems to have been limited connection, in the work undertaken, between the developers of the steam engine – Newcomen, Trevithick, Watt, and George Stephenson – and the scientists who preceded them, such as Hooke, Boyle, and Newton. The steam engine developers worked mainly by practical experimentation and development, rather than using theory and, if anything, their work prompted more investigation by the scientists, rather than vice versa.

In other cases, science clearly came before development. For example, maser (microwave amplification by stimulated emission of radiation) and laser (light amplification by stimulated emission of radiation) technology demonstrated in the 1950s by Charles Townes and others, and based ultimately on quantum theories developed by Max Planck in the early 1900s, was originally in the 'solution looking for a problem' category. Nowadays, of course, lasers can be found in a wide range of applications, from welding and metal-cutting machinery to hi-fi devices and then through to telecommunications using fibre-optics.

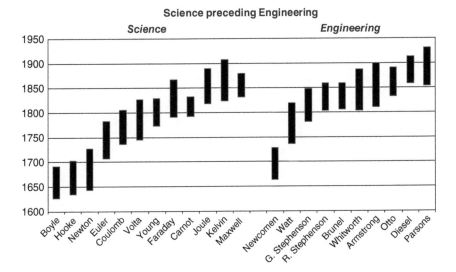

Figure 2.4 Life periods of great scientists and engineers.

2.6 Iteration as the Heart of the Process

As the preceding material suggests, an iterative loop is at the heart of the processes for developing technology and products. Sometimes described, somewhat pejoratively, as *test – break – fix*, it is really a learning cycle, as illustrated in Figure 2.5.

The cycle is used from the early stages of development – a sketch on the back of an envelope, a realisation that it's not quite right, followed by a redrawn sketch – to the later stages of sign-off when small adjustments are made prior to final release. Leonardo da Vinci was an early and prolific exponent of the engineering sketch, as illustrated in Figure 2.6.

The process could be compared to W. Edwards Deming's Plan/Do/Check/Act cycle, much used in manufacturing improvement. He, in turn, credited this

Figure 2.5 Learning cycle.

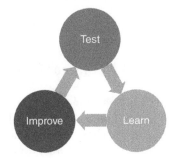

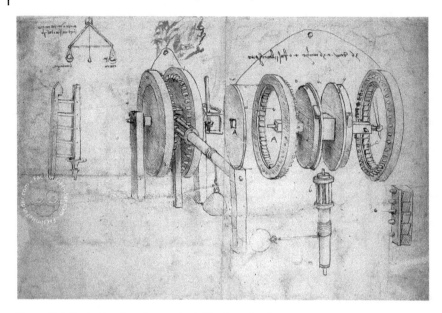

Figure 2.6 Illustration from Leonardo da Vinci notebook.

approach to Walter Shewhart, although its origins could be traced back to the seventeenth century as the 'scientific method': defining a hypothesis, experimentally testing it, then refining the hypothesis. Even the iterative model has its own multiple origins and cycle of improvement.

2.7 Impact of Low-Cost Computing

New technology has also affected scientific and engineering development. Readily accessible and low-cost computing methods have greatly enhanced the scope for more iterations of the learning cycle and hence potentially better results. Even something as simple as a spreadsheet, once set up, allows multiple 'what if?' studies to be undertaken. Going further, some complex models have the ability to be run repeatedly and automatically, homing in on the optimal solution. A feature of the Fourth Industrial Revolution will be more of this type of work.

2.8 A Nonlinear Process?

There is some debate as to whether technology and product development is a linear or nonlinear process. The real question here is whether or not the process can be characterised as a series of logical steps, which follow on from each other: do this and then do that, and so on. There is a degree of truth about both answers but it is certainly the case that you cannot draw a programme of events from the creation of a good idea to the launch of a successful product; innovation, unfortunately, is not that straightforward.

However, scientific discovery does tend to precede technology development. Design has to precede manufacture of prototypes. And detailed design does have to precede accurate cost estimation. There are many, many more examples of this logical progression, suggesting that there is at least a framework within which innovation, technology development, and product development must proceed.

But in the early stages of development, in particular, the exact destination and the route to get there will be uncertain, rather like exploring unmapped territory with unknown hills, valleys, rivers, and swamps. The exact application of a new technology may require some presentation of 'what-if's' to customers and then further ideas based on their responses. Quite often, the eventual application may be quite different from the inventor's original idea. For example, the piston-based steam engine developed by Thomas Newcomen and others in the early eighteenth century to pump water out of mines eventually led to the creation of rail networks. It is unlikely that this is what Newcomen originally had in mind.

These comments apply mainly to the early stages of development. Later-stage engineering development follows a more disciplined and structured sequence, with a map to follow. However, the map may not be 100% accurate and the occasional detour, hopefully rather short, may be necessary.

2.9   Multiple, Parallel Activities

To add to the complexity of technology and product development, there will usually be multiple development cycles running in parallel and interacting with each other. This is always the case with very complex products such as aircraft, automobiles, or trains, where teams of 100+ engineers are very common. For example, just take the engine compartments of cars or trucks, which are usually very congested areas where components compete for the available space. It is normal practice to build digital or physical mock-ups of these areas to flush out problems. As the designs for these areas progress, individual components interfere with each other and designs of individual components have to be iterated in parallel with each other to overcome these problems.

The same is true of systems, such as electrical power, air, or hydraulics. For example, as a design progresses, it might be found that the total electrical power requirements of the individual elements of a system exceed the output of the generator and a solution must be found either by reducing the power of elements or increasing the output of the generator. If the latter route is chosen, where does the mechanical power come from to create that extra electrical output?

The point to make is that multiple, parallel, or overlapping iterations have to be managed when a complicated product or technology is being developed.

This leads to the complexities of managing teams of engineers, with each person working on their own elements of a project. If there are two people in the team, there is then just one pairwise communication channel to manage. If there are three in the team, there are three pairwise channels; if there are 10 people, there are 45 channels; and if there are 100, there are 4950 channels in theory. In a large team, of course, everyone does not have to interact with absolutely everyone else but, as a first approximation, the number of communication channels goes up with the square of the number of team members (in a population of n, there are $n(n-1)/2$ channels). This can lead to a situation where adding people to a team slows down the work because the added burden of communication outweighs the additional effort – see Chapter 5 for more about resourcing.

2.10   Right First Time versus Iteration

The iterative process described above may seem at odds with the philosophy of 'right first time', which rightly pervades manufacturing thinking. The underlying argument of right first time, in a manufacturing context, is that it is always more efficient to avoid defects by preventative means than to allow defects to occur, to inspect the product to identify them, and then to correct them.

Where there is a clear and repetitive process for manufacturing an item, it clearly makes sense to ensure that it works as perfectly as possible and that a conforming product is always produced.

So is there a different interpretation of right first time in the technology and product development process? If the underlying purpose of a right-first-time approach is to prevent defects, then what constitutes a defect in this process? Given that the purpose of process is learning and improvement, then a failure in the process is a failure to understand the learning which has occurred during an iteration loop and/or a failure to act upon it. This points us towards a philosophy of thoroughly understanding every learning opportunity, whether from digital modelling or physical testing, analysing the results carefully and always acting upon them rather than sweeping uncomfortable truths under the carpet.

This philosophy is the one of the keys to quality management in technology and product development.

2.11   Lean Thinking Approach

These points can be developed further by applying 'lean thinking' philosophies, as described, e.g. in James P. Womack and Daniel T. Jones' book (Ref. 2.4) of the same title. This concentrates on the identification and elimination of waste. Lean thinking is based around five key points:

1) The distinction between valuable and wasteful activities in the context of creating value for the ultimate customer of the product or service
2) The identification of value streams – activities that create benefits for these end customers
3) The flow of valuable activities as a continuous, 'single-piece' process rather than a stop-start batch driven process
4) The concept of *pull*, that is, only doing tasks when they are needed and not before
5) The principle of continuous improvement and hence the pursuit of perfection

These principles are most readily applied to the physical processes of manufacturing. It is no exaggeration to say that they have led to a fundamental and revolutionary shift in manufacturing efficiency.

As the book above points out, the same principles can be applied in other fields, specifically mentioning order processing and product development.

In the case of technology and product development processes, wasted activities do clearly occur in this field and are always, in principle, capable of being prevented. Value streams may be less obvious to see but usually include cross-functional activities, involving departments outside the technology and engineering disciplines as well as suppliers. Because these value streams cut across organisational boundaries, they usually present the best opportunities for efficiency improvement but are the most difficult to tackle. *Pull* can be interpreted as doing the right things at the right time and not losing the synchronisation of cross-functional activities, not piling up information before it is needed or providing information late. Finally, the concept of continuous improvement can be applied to any business process. More is said about these points in Chapter 12, improving product development performance.

2.12   Cost of Problem Resolution

The learning cycle described above does have a cost in the sense that the analysis of experimental test results and subsequent improvement activities do take time and hence cost money. However, in the early stages of a technology, this may be no more than a 10-minute redraw of a back-of-the-envelope sketch. At the other end of the spectrum, if a problem occurs when a volume product such as a car is out in the marketplace, the cost of rectification can be in £ millions.

A rough estimate, illustrated on Figure 2.7, suggests a logarithmic scale of the cost of remedying a defect – it's shown in pounds (6 on the scale is one million) but it could equally be dollars or euros.

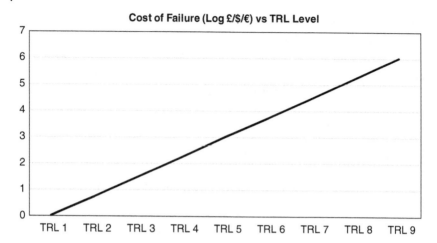

Figure 2.7 Cost of failure versus point in development cycle.

2.13 Risk versus Time

When development of a new technology is started, the risk of issues emerging once the technology is in production is very high. This is simply a function of the early stage of its development and the fact that, at that stage, there is a lot that could potentially go wrong in the future. One of the primary objectives of the subsequent technology and product development process is to reduce that risk so that, when it does go into service, the risk of problems is as close to zero as possible. In well-developed industries, this high standard can be achieved, although zero problems, i.e. perfection, will not.

This is shown graphically in Figure 2.8. The data are purely subjective and represent trajectories that might or might not be accurate.

This, then, raises the question of the extent to which risks can be identified and therefore anticipated – probably the most relevant and difficult question in the world of engineering.

The Johari window, Figure 2.9, or a variant of it, provides one framework for thinking, rather philosophically, about risk reduction in engineering. The original basic window is made up of a 2×2 grid, which lays out the relationship between personality traits known to the individual or known to others.

Developed by Joseph Luft and Harrington Ingham in 1955 (Ref. 2.5) as a means of mapping and understanding personality traits, a variant of it came to prominence through Donald Rumsfeld and his famous comments about 'unknown unknowns'.

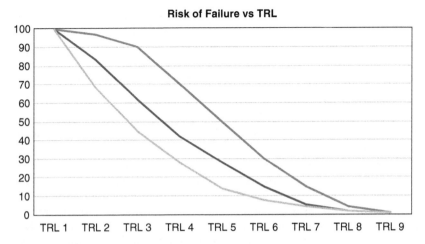

Figure 2.8 Percentage risk of failure versus stage of development.

	Known to self	Not known to self
Known to Others	Arena	Blind spot
Not Known to Others	Façade	Unknown

Figure 2.9 Johari window.

	Known to self	Not known to self
Known to Others	1) Consensus about the risks which could emerge – obvious	2) Risks that expert review might identify – experience
Not Known to Others	3) Risks that might be covered up or fail to be acknowledged – hidden	4) Danger zone – risks that come out of the blue

Figure 2.10 Engineering risk matrix.

It is also a good framework for thinking about engineering risk. The four categories of risk that the matrix in Figure 2.10 identifies could be thought of as follows:

1) *Obvious.* Those on which both the developer of the technology and outside parties agree and where therefore consensus can be easily reached about what development might identify and overcome those risks specifically.

2) *Experience.* Those that the developer may have missed but that peers and grey-haired engineers might identify from their experience.
3) *Hidden.* Those for which the signs are present but that the developer might be ignoring or dismissing, e.g. because test material is nonrepresentative.
4) *Danger zone.* Those that lie beyond the experience or expectation of all concerned.

A well-known example of a 'hidden' problem might be that which brought down the *Challenger* space shuttle in 1986. In this instance, danger signals in the form of data from earlier failures or problems with O-rings between sections of the solid rocket boosters were ignored. The disaster was investigated in great detail and it has become a case study in discussions of engineering safety and workplace ethics.

A less well-known example of the danger zone is the collapse of a railway bridge over the River Dee, in northwest England, which occurred in May 1847 – see Ref. 2.6. The design concept for the bridge was relatively simple and had been used extensively over the previous 15 years or so. The principal structural elements were two cast-iron girders forming the main spans of the bridge. Cast iron being weak in tension, the girders were asymmetric, with the bottom webs of the I-beam configuration having more cross-sectional area than the top webs. To provide still more load-carrying capacity, preloaded bars were also provided to pre-compress the bottom webs. These bars loaded the girders off-centre and therefore also introduced a twisting load as well as the intended compressive load. The bridge failed catastrophically as a train went over it, and the cause was eventually traced to torsional buckling instability of the main girders, a phenomenon which was not understood at that stage in engineering history. In some respects, this is a classic example of an *unknown unknown,* but arguably more caution could have been exercised, as the bridge span was substantially longer than anything attempted previously.

These points suggest the adoption of a disciplined or structured approach to dealing with (1) to (3), but the need for more creative methods of discovering the unknown unknowns and the application of caution when entering uncharted territory.

2.14   Creativity versus Risk Management

The previous sections of this chapter illustrate one of the most fundamental dilemmas of the engineering process.

On the one hand, it is an essentially creative process, developing new ideas to solve problems and to improve people's well-being. This is what attracts people into engineering and what provides the driving force to overcome problems. It is also the source of the innovation that drives business and economic growth in an industrial context. Jerome B. Wiesner, who was the thirteenth president of Massachusetts Institute of Technology, wrote, '*Technical and scientific work is usually fun. In fact, creative technical work provides much the same satisfaction that is obtained from painting, writing and composing or performing music'* (Ref. 2.7).

On the other hand, those new engineering solutions have to be reliable, robust, and not create harm or danger. In this respect, the world is becoming increasingly critical – for example, we expect to be safe when we fly, and the figures demonstrate a civil aircraft safety level of around 0.03 fatalities every 10^6 km travelled, a tenfold improvement in the last 70 years (Ref. 2.8).

Looking more widely, there is a strong expectation of safety and reliability in all the products or services that twenty-first century consumers buy, as illustrated in some further numerical performance data in Figure 2.11.

The exact numbers are not so important as the fact that, in developing a new technology or product, the bar is set very high in terms of the reliability and safety levels that must be achieved (see, e.g. Refs. 2.9 and 2.10).

We are also increasingly aware of the environmental impact of new technology and, of course, new technology can solve as well as create environmental problems. Hence, the engineering process must constantly be aware of risks

Industry	Reliability/ safety performance	Calculation basis	Source
Passenger aircraft	Fatality rate of 30 per 10^9 km of travel	Taken directly from survey	UK DfT survey for period 1990 – 2000
	Accident rate of 2.1 per 10^6 departures	Taken directly from report	ICAO safety report 2017
Automotive	Problem rate 1 per 625 operating hours (Breakdown rate will be several orders of magnitude lower.)	Best cars achieve c. 80 problems per hundred vehicles per year. Assume 500 operating hours per year.	JD Power UK Vehicle Dependability Study 2017 [2.7].
Process industry	Overall fatality rate of 1 in 1000 p.a.	'Broadly, a risk of 1 in 1000 p.a. is about the most that is ordinarily accepted under modern conditions for workers in the UK'.	UK Health & Safety Executive Publications [2.8].
	Single risk rate of 1 in 10^6 p.a.	Approximate limit above which ALARP (as low as reasonably practical) principles should be applied – see Chapter 7.	

Figure 2.11 Approximate safety and reliability performance.

and problems that must be identified and overcome before a new technology or product is launched.

In this book, Chapter 6 is particularly focused on the new ideas, whilst Chapter 7 is focused on risk management.

2.15 Early Detection of Problems

The obvious conclusion from the previous discussion is that ways should be found to detect problems at the earliest possible stage, before they become expensive. As will be described in Chapter 3, there is plenty of evidence that higher levels of early expenditure are associated with lower levels of subsequent cost and time overruns. This logic may seem obvious, but it is not always followed, and it is difficult to compensate for underdevelopment by applying more resources later.

There are five ways, however, of addressing the issue:

1) Build early mathematical models, mock-ups, and rig tests.
2) Plan the test programme at an early stage.
3) Work closely with materials, manufacturing, and supply experts.
4) Conduct expert reviews at an early stage.
5) Test ideas with customers, understanding and measuring their operating environment.

Thinking through what steps can be taken to detect problems at an early stage is probably the most cost-effective investment that can be made in a technology or product development programme.

2.16 Management of Change

Given that learning and iteration are at the heart of development processes, change management is a key activity, sometimes thought of as a purely bureaucratic process but actually something far more fundamental.

Change in this context refers essentially to modifications to the engineering information defining a product or technology. In the early stages of a project, it could just involve redrawing the sketches, or remaking the calculations that lie behind a new technology at this stage of development. More likely, the term will be used to cover modifications to the myriad of drawing, modelling, and textual information that is required to define a complex engineering product.

In the former stage, few people are involved and changes or improvements can be handled informally and by word of mouth. In the latter stage, many people are typically involved and they may be widely distributed geographically. Hence, formal change control processes are the norm, preserving the discipline but often seen as an obstacle to progress. Those tempted to short-cut might do well to study the 1981 Hyatt Regency walkway collapse in Kansas City – brought about by ill-considered design changes.

This is the essential dilemma of managing change – a process that is core to the development of new technologies and products but one that can be expensive and difficult. Two approaches can help, both of which come from the Japanese automotive industry in the 1970s and 1980s:

1) Encourage changes but make them at the earliest possible stage.
2) Make change management a consensual and cross-functional activity, rather than a bureaucratic process.

The first of these has already been covered. By identifying improvements at the earliest possible stage, the changes can be made with the least cost and frustration. This is also the stage when every effort should be made to minimise outright errors.

The second is where engineers, manufacturing specialists, and purchasing managers from all disciplines work towards the most effective solution. It does require an underpinning process, for example, to change documentation, to manage costs, to identify nonconforming material, and to agree the timing of implementation. If the organisation has been successful in flushing out improvements at an early stage, then these later-stage changes should be relatively minor. However, they are still important for improving the product, especially its manufacturability, and any efforts to eliminate them are very undesirable, if not unrealistic (complete design freezes never actually work). It is also undesirable to impose a very demanding change process with multiple authority levels, which just slows down the inevitable. It is far better to encourage cross-disciplinary working and to do this from the earliest possible stage.

2.17   Management of Learning

Changes, as described above, arise from learning about the technology or product under development. This will arise, typically, from analysis, manufacturing, or test work. Problems occur, and solutions may be found immediately or after further work. This learning is among the most valuable intellectual property that derives from development work. It is vital that it is not lost, and an effective problem or learning recording system should be an integral part of the development process. It does not need to be complicated and could simply include:

- A description of the problem encountered
- A record or analysis of the details of the problem
- An assessment of the root cause of the problem
- A note of the potential solution
- A record of the problem having been closed out in subsequent phases of work

As with the change process, learning processes can be handled informally or by word-of-mouth in the early stages of projects. Once the landscape broadens, a more formal process ensures that learning is not lost. For example, if early prototype manufacture is put out to subcontractors for parts manufacture, they will undoubtedly find problems, sometimes quite minor, and these opportunities for improvement should be captured. An important part of the latter stages of new product programmes is then working through all the records of all the problems that have been logged and ensuring that they have been dealt with.

2.18   Governance of the Process

The technology and product development processes described above do not lend themselves to detailed control and supervision, unlike, e.g. a manufacturing processes where items can be measured as they are made and rejected if nonconforming. In the engineering or technology development process, defects will not appear until much later, when the technology is actually being used in service, which could be a number of years after the defect is created. At the same time, the processes are critical to the long-term health of organisations as the source of future customers, markets, and revenues.

Company management must therefore provide oversight to the process but cannot be expected to run it in detail. Strategic oversight can be provided through the following:

• Providing a climate within the company that encourages innovation
• Deciding, by screening of early-stage technologies, which have potential future value and should be taken further
• Similarly deciding which technologies or products should be launched into a commercialisation phase, which could involve a major financial commitment
• Participating in formal reviews of products, via a stage-gate process, for example, as they progress through the commercialisation phase
• Applying business analysis to the decisions noted above
• Undertaking business reviews of launched products in the marketplace to compare their performance with that planned

The points noted above are essentially concerned with an organisation's technology and product policy, which are clearly the domain of senior management and a critically important aspect of company strategy. For more discussion of this subject, see Ref. 2.11.

2.19   Formal Quality Management Systems

The points just noted relate to the policy management, by senior management, of technology and product development work. To complement this approach at the more detailed working level, formal quality management systems must also be applied. This is especially relevant when the product has safety-critical elements to it, although it is just as important to good practice in all applications.

The engineering disciplines are responsible for generating and maintaining, for the organisation as a whole, data in the form of drawings, bills of material, specifications, test codes, and so on. These are important information assets upon which the whole enterprise depends in terms of accuracy, completeness, and timeliness. Without going into the structure and detail of a formal quality management system, the areas which need to be covered include:

- Responsibilities and accountabilities
- Identification and change control of drawings, documents, and specifications
- Management of bills of material
- Maintenance and calibration of test facilities and instrumentation
- Recording of results of analysis and test work
- Identification and traceability of test materials
- Recording and close-out of risks and problems
- Maintenance of data integrity and security
- Audit of system performance

In an early-stage company, these disciplines may seem unnecessary, but ignoring them tempts fate in the form of lost data, incorrect results, or misidentification of hardware. An ISO9001-type system is vital from quite early stages of development. Large companies will already have such systems in place; smaller companies often delay the inevitable, to their detriment. It should be noted that these smaller companies will, sooner or later, find that they will be unable to do business without such certification, which is a mandatory requirement in the purchasing policies of many companies.

2.20 Concluding Points

The conclusion from this chapter is that there is an engineering, technology, and product development process. However, the process can be rather unstructured in its early stages, in contrast to the much more disciplined later stages. It is not a highly repetitive process: each project runs for a long time and is rarely repeated identically. However, it is characterised by a series of learning cycles. Multiple cycles often run parallel to each other, creating a highly interactive, as well as iterative, process.

As a fundamental principle, the more investment that can be made in the early cycles, and the more they can be cross-disciplinary, the better the results of the later stages are likely to be. This, then, leads to the next chapter, which examines how the status and maturity of a new development can be assessed.

Perhaps the most perceptive comment about the engineering process comes from Appendix F of the *Challenger* disaster report (Ref. 2.12), an appendix written as a personal statement by Richard P. Feynman – a Nobel Prize winner in physics: *'For a successful technology'*, he concluded, *'reality must take precedence over public relations, for nature cannot be fooled'*. In other words, you can't fool or short-cut the development of technology (although you can improve its efficiency).

References

These three references provide information about R&D from an economic perspective:

2.1 2016 global R&D funding forecast *R&D Magazine, Advantage Business Media*. 2016.

2.2 Global Innovation Index *The Economist*, 2015.

2.3 Measuring corporate R&D returns - Bronwyn H. Hall, University of California at Berkeley and Maastricht University, Jacques Mairesse, CREST-ENSAE and UNU-MERIT

Three references provide useful material about risks, problems, and waste in engineering systems:

2.4 Petroski, H. (1994.). *Design Paradigms – Case Histories of Error & Judgement in Engineering*. Cambridge: Cambridge University Press.

2.5 Luft, J. and Ingham, H. 1955. The Johari window, a graphic model of interpersonal awareness. Proceedings of the Western Training Laboratory in Group Development. Los Angeles, University of California, Los Angeles

2.6 Womack, J.P. and Daniel, T.J. (1996). *Lean Thinking*. New York: Simon & Schuster Inc.

Thoughts from nineteenth-century leaders in the field of engineering, including Wiesner, can be found in this compilation, still useful more than 50 years later.

2.7 Love, A. and Chambers, J.S. (1966). *Listen to Leaders in Engineering*. Philadelphia: David McKay.

Information about aircraft, automobile, and process plant reliability and safety is available at

2.8 ICAO Safety Report 2017 – United Nations International Civil Aviation Organization (ICAO) https://www.icao.int/safety/Documents/ICAO_SR_2017_18072017.pdf

2.9 JD Power UK Vehicle Dependability Survey 2017 (VDS).

2.10 Rick, J., Evans, C., and Barkworth, R. (2004). *Evaluation of Reducing Risks, Protecting People*. Institute of Employment Studies for the UK Health & Safety Executive, RR279.

More can be read about the overall governance of product development processes at

2.11 Haines, S. (2013). Strategies to improve NPD governance. In: *PDMA Handbook of New Product Development*, 3e (ed. K.B. Kahn). Hoboken, NJ: Wiley.

The final comments come from the report into the *Challenger* disaster

2.12 Rogers Commission Report, Volume 2, Appendix F – Personal Observations on Reliability of Shuttle, NASA – Richard P. Feynman, 1986.

3

Evaluating the Maturity of Developing Technology

3.1 Background

Amongst the most fundamental questions to ask when developing a new technology or product is: 'How far have I got and what remains to be done?' If the answers to such questions were easy, then the past would not be littered with examples of technologies that have gone wrong.

Every engineer has his favourite stories, either from history or from his own experiences.

Scandinavians like to quote the example of the warship *Vasa*, which listed and sank immediately after being launched in August 1628. The ship's builders had been too afraid to tell the Swedish king that his ideas wouldn't work. And there is a lesson in this tale even for today's engineering companies – there is still a fear of giving bad news.

In the United Kingdom, an oft-quoted example is the Advanced Passenger Train (APT). The ambition in the 1980s was to develop a tilting passenger train that could negotiate the 'curvaceous' West Coast Main Line and reduce journey times from London to Glasgow, which it was well capable of doing (in fact, at the time of writing, the APT still holds the record of 3 hours 55 minutes for the northbound journey). However, the technology was developed in the public eye and involved fare-paying passengers – something that no right-minded engineer would ever recommend. Not surprisingly, the problems that arose on the development programme, which were quite typical for a technology still maturing, became a source of ridicule and embarrassment. The project was cancelled. Today, tilting trains run very successfully on this route using technology based on the APT's but made by someone else.

The IT world has provided additional evidence. Even Apple, the doyen of domestic computing, introduced and quietly withdrew, for example, the Apple Newton, which achieved some notoriety at the hands of the strip cartoonists.

Could these problems have been avoided if there had been a better understanding of the maturity of the technologies in question? There are no guarantees of a magic cure, but surely there would have been more chance of avoiding

Managing Technology and Product Development Programmes: A Framework for Success,
First Edition. Peter Flinn.
© 2019 John Wiley & Sons Ltd. Published 2019 by John Wiley & Sons Ltd.

embarrassment if there had been an objective way of measuring the readiness of these technologies and then acting on it, thus avoiding the loss of reputation that always accompanies publicly obvious failures.

3.2 Origins of Technology Readiness Measurement

This form of problem faced the National Aeronautics and Space Administration (NASA) in the 1970s. The US space program had a large portfolio of developing technologies that might be used on future missions. But which of them could be made to work, and by when? Historical cost analyses had shown that there was a direct link between the success of a technology and the amount (of dollars) spent on its early-stage development – the more the expenditure at that stage, the less likelihood there was of cost and timescale overruns. Most engineers would regard this as a statement of the obvious, but NASA cost engineers demonstrated the point and put some numbers to it.

The question was obviously crucial to planning of future programmes, especially manned space flights where there is no room for error. According to Donna Shirley (Ref. 3.1), manager of Mars Exploration at the Jet Propulsion Laboratory, the business of technology readiness levels (TRLs) got started at NASA because of a 'guy named Werner Gruel, who was a NASA cost analyst.' Gruel 'had this great curve that showed that, if you spent less than 5% of the project cost before you made a commitment to the technology, it would

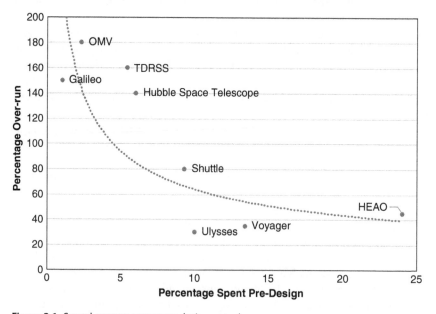

Figure 3.1 Spend overrun versus predesign spend.

overrun. If you spent 10% of the total cost before that commitment, it would not overrun'. The curve is reproduced in Figure 3.1.

The approach to addressing this issue and taking it forward is credited to NASA researcher Stanley Sadin and two of his colleagues. In 1974, he developed a seven-step system for estimating the readiness of new technologies. It was formally defined in a published report in 1989. It was then extended and used more publicly by NASA from the early 1990s as a nine-step system, running from initial concept to validation by operation in space – see Ref. 3.2–3.5.

This approach has grown in application over the subsequent 25 or so years by an expansion of use by both public and private organisations.

3.3   Purpose of Technology Maturity Assessment

Since its original development to support the technology planning of space programmes, a range of uses has been found for technology readiness assessment methods:

- Understanding the maturity of individual projects within a portfolio of technology development projects. This helps relatively complex organisations to understand what they have in their development pipeline and the timescales within which a particular technology is likely to mature.
- Evaluating the status of a specific technology, or product development, programme prior to deciding whether to invest resources in further development or to launch into a live production programme. Many larger companies have institutionalised this approach as a 'stage-gate' system where formal sign-off is required before that technology can be taken further – see Chapter 4.
- Judging whether a product or technology is worthy of external investment. This is particularly relevant to angel or venture capital-type investment organisations who often struggle to understand the maturity of early-stage companies requesting funding.
- Managing the procurement or acquisition of technology by (public) procurement bodies such as those running defence programmes. Here the idea is to make sure that what has been promised is being delivered and that the procuring organisation is not throwing good money after bad.

3.4   Users of Technology Maturity Assessment

Within this context, technology readiness assessment is, or could be, used by quite a wide range of organisations, including:

- Space and defence system companies, the original users
- Public procurement/technology acquisition bodies, such as national defence procurement organisations

- Engineering companies and complex system integrators, e.g. aircraft manufacturers, jet engine manufacturers, car and commercial vehicle manufacturers, railway rolling stock manufacturers, oil and gas companies
- Nuclear industry, including those involved in decommissioning
- Investors such as venture capital and 'angel' investors
- Pharmaceutical or biomedical products, e.g. US Department of Health & Human Services

The principles of its use can also be extended into other fields such a software development.

3.5   What Is Technology Maturity?

It may seem like a statement of the obvious, but a mature technology is one that works reliably when in the customer's hands and can be manufactured consistently at the appropriate cost without having to use specialised or difficult methods. It is likely to be used in a range of applications by a number of companies.

An immature or underdeveloped technology is the opposite of these and will at best cause frustration to the end user and at worst be downright dangerous. In the early stages of developing a technology, where technology is considered to be the application of science into something useful, immaturity translates into uncertainty and risk, and then into cost and time overruns. Arguably, the most crucial question in applying a new technology is whether it has reached a sufficient level of maturity to be inserted into a planned new product programme where a commitment will be made to 'go live' at a certain date, with all the associated costs and investment as well as the exposure to end customers.

3.6   Technology Readiness Level (TRL) Structure

The structure developed originally by NASA has stood the test of time and has been developed by different industries to suit their needs. Three of these structures are summarised in Figure 3.2. Alongside the NASA system itself are the definitions for each TRL used by the European Union in the Horizon 2020 research programme. There is then a 10-step system developed by the UK automotive industry, using the language, processes, and practices of the automotive industry – Ref. 3.6. These are presented as illustrations. Further examples can be found from a wide range of industries such as energy and rail.

TRL	NASA[*]	EU H2020 Definitions	UK Automotive, slightly simplified
1	Basic principles observed and reported	Basic principles observed	Basic Principles observed and reported. Scientific research undertaken and beginning to be translated into applied research and development. Paper studies and scientific experiments taken place and performance predicted.
2	Technology concept and/or application formulated	Technology concept formulated	Speculative applications identified. Exploration into key principles ongoing. Application specific simulations or experiments undertaken and performance predictions refined.
3	Analytical and experimental critical function and/or characteristic proof of concept	Experimental proof of concept	Analytical and experimental assessments identified critical functionality and/or characteristics and physically validated predictions of separate elements of the technology or components not yet integrated or representative. Performance investigation using analytical experimentation and/or simulations underway.
4	Component and/or breadboard validation in laboratory environment.	Technology validated in laboratory	Technology component and/or basic subsystem validated in laboratory or test house environment. The basic concept observed in other industry sectors. Requirements and interactions with relevant vehicle systems determined.
5	Component and/or Breadboard validation in relevant environment.	Technology validated in relevant environment	Technology component and/or basic subsystem validated in relevant environment, potentially through a mule or adapted current production vehicle. Basic technological components integrated with reasonably realistic supporting elements so that the technology can be tested with equipment that can simulate and validate all system specifications within a laboratory, test house or test track setting with integrated components Design rules established. Performance results demonstrate the viability of the technology and confidence to select it for a new vehicle programme

Figure 3.2 Definition of TRL levels.

TRL	NASA[*]	EU H2020 Definitions	UK Automotive, slightly simplified
6	System/sub-system model or prototype demonstration in an operational environment.	Technology demonstrated in relevant environment (industrially relevant environment in the case of key enabling technologies)	A model or prototype of the technology system or subsystem has been demonstrated as part of a vehicle that can simulate and validate all system specifications within a test house, test track or similar operational environment. Performance results validate the technology's viability for a specific vehicle class.
7	System prototype demonstration in an operational environment.	System prototype demonstration in operational environment	Multiple prototypes demonstrated in an operational, on-vehicle environment. The technology performs as required. Limit testing and ultimate performance characteristics now determined. Technology is suitable to be incorporated into specific vehicle platform development programmes.
8	Actual system completed and "flight qualified" through test and demonstration.	System complete and qualified	Test and demonstration phases have completed to customer's satisfaction. Technology proven to work in its final form and under expected conditions
9	Actual system flight proven through successful mission operations.	Actual system proven in operational environment	Actual technology system has been qualified through operational experience. Technology applied in its final form and under real-world conditions.
10			Technology is successfully in service in multiple application forms, vehicle platforms and geographic regions. In-service and life-time warranty data available, confirming actual market life, time performance and reliability

[*]https://www.nasa.gov/pdf/458490main_TRL_Definitions.pdf

Figure 3.2 (Continued)

3.7   Phases of Technology Readiness

The nine TRL levels can be divided into three distinct phases:

- TRL 1–3, which is usually carried out in a university or corporate research laboratory

- TRL 4–6, which is harder to categorise and which could be done in either a university, a research organisation or an industrial test house
- TRL 7–9, which is very much in the industrial or commercial domain

The first phase establishes the basic principles of the technology – that it can be made to work. It is likely to require practical laboratory development by highly skilled staff and using supporting calculations or mathematical simulations. The emphasis will be very much on making the technology function through a lot of iteration, rework, skill, and ingenuity. The dependability of the technology might be quite limited; we are all familiar with the laboratory demonstrations that worked successfully an hour ago but that fail when shown to an important visitor. There will, however, be much learning about the critical factors in making the technology a success. Finally, there will be also be some understanding of the likely market applications of the technology, how it might be manufactured, and roughly what it might cost.

The intermediate TRL 4–6 phase has the aim of making the technology more robust and predictable so it could then be deployed, with confidence, into a detailed product development and manufacturing launch phase. Testing will become increasingly realistic in terms of the scale of any prototypes and will be conducted in an environment approximating the true operational situation. This could require significant facilities and supporting calculations. Simulations will be very detailed. Full design for manufacture (DfM) studies should be undertaken so the design does not have to be amended subsequently to make it capable of production (many technologies effectively regress if the DfM phase is left too late). This work will also flush out manufacturing costs and highlight any need for further development to hit realistic cost targets.

The decision to move beyond TRL 6 is one of the most critical in the product development cycle, as it leads to substantial expenditure where the scope to respond to immature technology is much more limited, in timescale terms, and is increasingly expensive. However, the TRL 4–6 phase is also one where there is pressure to short-cut and compromise because the benefits of the phase are often only apparent later, or conversely, problems arise that could have been ironed out earlier.

The final, TRL 7–9 phase is the easiest to understand and by far the most expensive. Full and detailed product definition in the form of computer-aided design (CAD) models, drawings, bills of material, and specifications are completed. Calculations and modelling are finished. Prototype and pre-production programmes are undertaken to demonstrate the correct functioning of the technology and its durability, often using accelerated test regimes. Legally required certification is achieved and the full production system is launched.

Many companies and industries have detailed sign-off requirements to ensure that a robust and legal product is produced. They may also have stage-gate decision-making processes which subdivide the progress through the different stages of maturity.

Progress through the TRL phases described above is characterised as follows:

- An increasing level of detail in the definition of the product or the technology
- Increasingly rigorous and representative (and costly) testing, test environment and supporting analysis
- Increasingly realistic test items, produced by increasingly representative methods
- Reducing level of skill in the production of development and test material

At the same time as developing the maturity of technology and products, the corresponding manufacturing processes must be developed and thought through. Without this, the new product may not be capable of efficient production.

3.8 The 'Valley of Death'

No discussion of technology maturity development is complete without touching on the concept of 'the valley of death'. This term is frequently used in both the public and private arenas in the context of moving early-stage companies from concept demonstration to financial viability. It is an imprecise and evocative term used to describe a period where the majority of start-ups come to grief and where public funding of earlier technology research may therefore be seen as wasted.

As indicated above, this phase or phases of work include the period when the technology itself undergoes thorough development, beyond concept demonstration, and is then followed by detailed engineering and investment in manufacturing launch. A cash flow profile might typically look like Figure 3.3 where

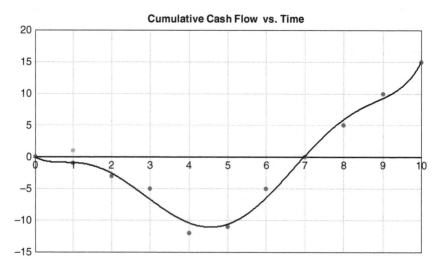

Figure 3.3 Start-up cash-flow profile.

there are four to five periods of cash outflow before sales revenue starts to come in.

This profile applies whether the work is undertaken by a start-up or an established corporation. However, in the latter case, the nature of the work will be better understood and experience will provide a better indication of the costs and timescales involved. There will probably also be a better understanding of the market potential and hence whether an adequate return can be earned from the investment.

In the case of start-up companies, the opposite points apply. There is likely to be more of a tendency to underestimate what is required, in terms of both time and money, something guaranteed to annoy investors. This can then lead to shortages of funds and attempts to short-cut, leading to underdeveloped products.

Can this be avoided? Stating the obvious, companies in this situation need robust marketing plans, realistic development plans and costs, and ideally early sales of trial units to supportive customers. More is said on the topic of early-stage funding in Chapter 6.

3.9 Manufacturing Readiness Level (MRL) Structure

Moving back to readiness scales, the technology readiness concept has led to a proliferation of offshoot maturity scales in manufacturing, supply chains, and software, for example. The most important and relevant for technology development is manufacturing readiness, which can be applied either to a new manufacturing technology or to the manufacture of a new product technology using existing manufacturing methods. Figure 3.4 summarises the principles of the manufacturing readiness level (MRL) system used by the US Departments of Defence and of Energy, which uses similar language to the NASA TRL system. Alongside it is the scale developed by the UK automotive industry, which uses the language and activities of this particular industry. One of the key points about the MRL scales is the breadth of activities they cover. As well as manufacturing itself, there is coverage of materials availability, the supporting industrial base including supply chains, the business case, skills availability, and supply schedules. The term *manufacturing readiness* might well be replaced by *business readiness*.

3.10 Progressing through the Scales – Some Practical Points

The principles outlined above do require some effort to be translated into practical use. The language used in the TRL or MRL definitions is quite general and its origins in the space and defence industries do come through, even when adapted to other industries. Figures 3.5 and 3.6 outline in more practical terms what might be undertaken or completed in each stage of the TRL/MRL journey in a way that relates to the language of general company management.

MRL	Specified by DoDI 5000.02	UK Automotive, slightly simplified
1	Basic manufacturing implications identified	Basic manufacturing Implications identified. Materials for manufacturing characterised and assessed
2	Manufacturing concepts identified	Manufacturing concepts and feasibility determined and processes identified. Producibility assessments underway including advanced design for manufacturing considerations.
3	Manufacturing proof of concept developed	Manufacturing proof-of-concept developed. Analytical or laboratory experiments validate paper studies. Experimental hardware or processes created, but not yet integrated or representative. Materials and/or processes characterised for manufacturability and availability. Initial manufacturing cost projections made. Supply chain requirements determined.
4	Capability to produce the technology in a lab environment	Capability exists to produce the technology in a laboratory or prototype environment. Series production requirements, such as in manufacturing technology development, identified. Processes to ensure manufacturability, producibility and quality in place and sufficient to produce demonstrators. Manufacturing risks identified for prototype build. Cost drivers confirmed. Design concepts optimised for production. APQP processes scoped and initiated.
5	Capability to produce prototype components in a production relevant environment.	Capability exists to produce prototype components in a production relevant environment. Critical technologies and components identified. Prototype materials, tooling and test equipment, as well as personnel skills demonstrated with components in a production relevant environment. FMEA and DFMA have been initiated.
6	Capability to produce a prototype system or subsystem in a production relevant environment.	Capability exists to produce integrated system or subsystem in a production relevant environment. Majority of manufacturing processes defined and characterised. Preliminary design of critical components completed. Prototype materials, tooling and test equipment, as well as personnel skills demonstrated on subsystems/ systems in a production relevant environment. Detailed cost analyses include design trades. Cost targets allocated and approved as viable. Producibility considerations shaping system development plans. Long lead and key supply chain elements identified.

Figure 3.4 Definition of MRL levels.

MRL	Specified by DoDI 5000.02	UK Automotive, slightly simplified
7	Capability to produce systems, subsystems or components in a production representative environment.	Capability exists to produce systems, subsystems or components in a production representative environment. Material specifications approved. Materials available to meet planned pilot line build schedule. Pilot line capability demonstrated including run at rate capability. Unit cost reduction efforts underway. Supply chain and supplier Quality Assurances assessed. Long lead procurement plans in place. Production tooling and test equipment design & development initiated FMEA and DFMA completed.
8	Pilot line capability demonstrated. Ready to begin low rate production.	Initial production underway. Manufacturing and quality processes and procedures proven in production environment. An early supply chain established and stable. Manufacturing processes validated.
9	Low Rate Production demonstrated. Capability in place to begin Full Rate Production.	Full/volume rate production capability has been demonstrated. Major system design features stable and proven in test and evaluation. Materials available to meet planned rate production schedules. Manufacturing processes and procedures established and controlled to three-sigma or some other appropriate quality level to meet design characteristic tolerances in a low rate production environment. Manufacturing control processes validated. Actual cost model developed for full rate production.
10	Full Rate Production demonstrated and lean production practices in place.	Full Rate Production demonstrated. Lean production practices in place and continuous process improvements on-going. Engineering/design changes limited to quality and cost improvements. System, components or other items in rate production and meet all engineering, performance, quality and reliability requirements. All materials, manufacturing processes and procedures, inspection and test equipment in production and controlled to six-sigma or some other appropriate quality level. Unit costs at target levels and applicable to multiple markets. The manufacturing capability is globally deployable.

Figure 3.4 (Continued)

TRL LEVEL

	1	2	3	4	5	6	7	8	9
Product definition/description									
Idea	*	*	*	*	*	*	*	*	*
Sketches & narrative description		*	*	*	*	*	*	*	*
Schematics, models, general arrangements				*	*	*	*	*	*
Limited change control in place				*	*	*	*	*	*
Design rules & operating environment defined					*	*	*	*	*
Formal change control in place						*	*	*	*
Detailed CAD models, specs, BoM's							*	*	*
Full production info								*	*
Form of physical realisation									
Hand-made/lash-up	*	*	*	*	*	*	*	*	*
Fully representative but not off tools							*	*	*
Off tools								*	*
Off tools to final production process									*
Quantities produced									
One-off	*	*	*	*	*	*	*	*	*
4 to 10				*	*	*	*	*	*
10 to 100							*	*	*
Production quantities									*
Testing & validation									
Calculation	*	*	*	*	*	*	*	*	*
Computer models			*	*	*	*	*	*	*
Lab test,			*	*	*	*	*	*	*
DFM, FMEA …					*	*	*	*	*
Accelerated/extreme testing						*	*	*	*
Real environment testing							*	*	*
Real environment testing with end users								*	*
Regulatory approval									
Approach to regulatory approval defined						*	*	*	*
Full approval									*

Figure 3.5 Achievement of TRL levels.

	MRL LEVEL								
	1	2	3	4	5	6	7	8	9
Process definition/description									
General concept proposed with justification	*	*	*	*	*	*	*	*	*
Specific process and application defined			*	*	*	*	*	*	*
Process fundamentals fully understood				*	*	*	*	*	*
Production process defined						*	*	*	*
Production process and all supporting data fully defined								*	*
Process capability validation									
Simple test work		*	*	*	*	*	*	*	*
Repeatable process proved by production of specific part(s)			*	*	*	*	*	*	*
Tests confirm process rate and likely production/capital costs				*	*	*	*	*	*
Process validated on production equipment & tooling					*	*	*	*	*
Process capability statistically validated						*	*	*	*
Production facilities & tooling used								*	*
Production facilities & tooling used in routine production									*
Quantity produced									
One-off	*	*	*	*	*	*	*	*	*
4 to 10			*	*	*	*	*	*	*
10 to 100					*	*	*	*	*
Production quantities									*
Skill required to produce									
Highly skilled and versatile		*	*	*	*	*	*	*	*
Skilled operator					*	*	*	*	*
Trained production operator						*	*	*	*
Production facilities									
Facility & process understood in outline form		*	*	*	*	*	*	*	*
Core process and facilities defined			*	*	*	*	*	*	*
Approach and budget defined				*	*	*	*	*	*
Production facility modelled					*	*	*	*	*
Production process trialled						*	*	*	*
Production facilities installed								*	*
Facilities proven at production rate									*

Figure 3.6 Achievement of MRL levels.

Cost management

Broad idea of cost range

Overall cost target & business case

Item-by-item targets

Validated by quotes

Costs validated in production

Business case

Outline business case defined

Full business case agreed ahead of MRL 7- 9 commitment

Full business case confirmed with real data

Timing plans

Programme of work to achieve MRL4

Approximate timescale/process to MRL9 understood

Programme to MRL9 defined

Firm SoP date & plan to reach MRL9 defined

Supply chain management

Make vs buy study completed

Critical suppliers selected

All suppliers selected

Initial schedules in place

Routine MRP scheduling in place

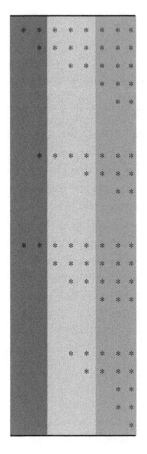

Figure 3.6 (Continued)

3.11   International Standards

In terms of officially recognised standards relating to this topic, the only one which currently exists is ISO16290:2013 – 'Space Systems – Definition of the TRLs and their criteria of assessment' – Ref. 3.7. It was produced by ISO Technical Committee 20 – TC 20 – Aircraft and space vehicles, Sub-Committee SC14, Space systems and operations and covers 11 pages.

The standard provides useful information:

- Definition of nine TRL levels
- Examples of the readiness level of various technologies at different points of time in the past
- Work expected to have been achieved on completion of each TRL level

The language used in the standard is very much that of the space industry and, although it is explained well through definition of key terms (e.g. 'breadboard', 'element' and 'relevant environment'), it would be more difficult to use directly in other industries using different terminology and processes. The less official documents produced by other industries, or indeed other individual companies, show that the principles of the system can be adapted to a wide range of circumstances.

3.12 Assessment of TRL and MRL Levels

TRL and MRL assessment can be carried out in individual companies on their products. When a company decides to make use of a TRL or MRL system, it is up to them to decide how to do so and how to build it into their formal processes. Normal practice is to have a documented set of definitions and criteria that describe the process to be used in the company and what is required to pass each TRL or MRL point.

Starting with technology readiness, questions such as these should be asked:

- What level of detail in product definition exists, in terms of drawings, specifications, bills of material, and so on?
- What form of physical realisation has been achieved, and how realistic is it compared with the final product?
- What calculation, modelling, and simulation work has been completed to verify the performance of the design, and what results has this work shown?
- What physical test work has been undertaken, in what quantities, and in how realistic a test environment? What results has this shown?
- Has intellectual property been protected?
- Is there a programme for regulatory approval, and how far has it progressed?
- To what extent has manufacturing feasibility been assessed, and how far has design-for-manufacture progressed?

Figure 3.7 is an example of a worksheet that could be used to assess whether a technology had reached, in this instance, TRL 3.

Manufacturing readiness can then be questioned in the same way:

- Is there a clear understanding of the manufacturing technology and processes needed to produce the planned product?
- Is there an understanding of the planned manufacturing volumes and manufacturing costs as a function of time that link to a viable business case and an introduction programme?
- Have the manufacturing processes and quality management arrangements been defined and modelled?

Technology Readiness Level - TRL3		TRL achieved;	Yes/No
TRL3 – Basic concept of the new technology is shown to be viable (proof of concept). On the basis of this, further investment in the next, and more expensive, stages of development might be expected by either the organisation doing the development or by an external investor…		Overall Comments	
Area	**Supporting Information expected at this stage**	**Pass/Fail**	**Issues/comments**
Concept design	The design of the solution will be complete, as an overall system but not the details. This will probably be in the form of a CAD model plus coding of any embedded software.		
Simulation & modelling	Substantial simulation and modelling will have been undertaken to prove the performance characteristics of the system.		
Practical development	This will be supported by practical test work, in a laboratory environment, of a complete system, albeit using relatively crude representations of the final items. This demonstration system is unlikely to work reliably, and could be quite temperamental, but it will show that the concept can be made to work and can achieve something approaching the required level of performance.		
Marketing evidence	The development work will also be supported by an initial QFD-type (quality function deployment) analysis, linking the product in detail to the market need that has been identified.		
Risk identification	FMEA or similar methodologies will also be started to identify areas critical to the reliability of the solution.		
IP protection	The need for formal IP protection will be identified and may have been sought.		
Manufacturability	A manufacturing approach will be defined and documented, suited to the volumes of the market identified.		
Financial viability	A basic assessment will be available to demonstrate financial viability.		

Figure 3.7 Example of TRL assessment sheet.

- Are the required materials and manufacturing infrastructure available, either within the company or outside it in the supply chain, in sufficient capacity to support the planned volumes?
- Is there a costed programme of investment in facilities and tooling?
- Have the manpower and skill requirements been defined?

These questions could form the basis of a review process that might be undertaken by independent staff in the company or through formal review meetings, also involving independent or experienced staff. To support this process, there is likely to be some form of check-sheet system which records the outcome of the review and defines the TRL or MRL status of the technology or product. Reviews should be evidence-based in the sense that they look directly at the available evidence and do not rely on second-hand information or opinion. (The promoters of new technology usually overestimate its readiness and brush aside known weaknesses.)

There are also publicly available spreadsheet-based calculators, such as that published by the American Air Force Research Laboratory – Ref. 3.8. As might be expected, it is aimed at aerospace and defence users, but the principles could be used across a wider range of industries. In this example, there is one sheet of questions, 10–20 in number, for each TRL level. These can be ticked off or a completion percentage estimated. The spreadsheet then calculates the resulting TRL outcome.

One point of debate occurs when a technology or product passes most of the criteria at a specific review point but still has some deficiencies. The correct argument is that these deficiencies should be corrected before a 'pass' can be awarded. Otherwise, the company misleads itself about where it has reached and is storing up problems for the future. Hence, the TRL or MRL level of a system is, in effect, dictated by the lowest ranking, or weakest, element of that system.

Figure 3.8 shows the typical outcome of a real TRL assessment, in this instance a new development in the field of renewable energy. In this example, the market was well understood, there was a lot of engineering detail with some degree of IP protection, prototypes were well made but constructed very laboriously, test work had been completed to demonstrate the concept but a more thorough programme had still to be undertaken, and risk and manufacturing work were still at early stages. Hence, although the design might be considered as being at TRL5, overall, the project was still at TRL3 and the various elements of it were out of synchronisation with each other. In particular, there was a concern that when a satisfactory method of manufacture had been found, then substantial redesign and retesting might be needed.

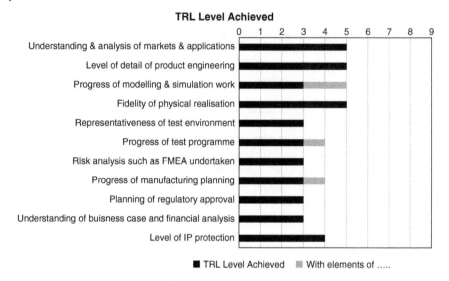

Figure 3.8 Example of results from a TRL assessment.

3.13 Synchronising Technology and Manufacturing Maturity

One of the key principles put forward in this book is that technology development can only be successful if the associated manufacturing development is carried out closely in parallel. In TRL and MRL terms, this means that MRL should never lag TRL by more than a couple of points (the best or acceptable lag is a matter of some debate).

This ideal relationship is illustrated in Figure 3.9. The graph also illustrates what can happen if technology development runs too far ahead of the underpinning manufacturing work. Typically, it is discovered that the design, which has been detailed and tested, cannot be made economically, needs to be modified to facilitate manufacture, or needs to be made from a somewhat different material. Redesign is then necessary, resulting in the TRL level regressing and development work having to be repeated. Interpersonal factors can often then intervene, recrimination sets in, and the project never quite recovers.

The key point is always to ensure early assessment of manufacturing, supply chain, and business parameters in the low-TRL stages of a new development (see, e.g. Ref. 3.9).

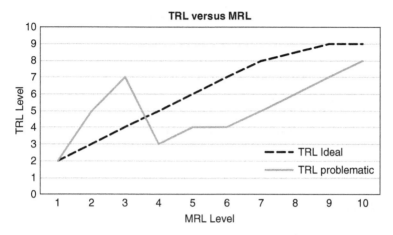

Figure 3.9 TRL versus MRL relationship.

3.14 Limitations of Technology Maturity Assessment

One point to bear in mind when conducting maturity assessments is the consistency of TRL definitions. There has been much debate about how the various TRL and MRL levels should be defined. Different industries and different companies have taken their own approaches. Hence, the TRL or MRL declared by one organisation may not be directly comparable with one declared by another. The difference, however, is only likely to be ±1 TRL or MRL level.

However, the most obvious drawback is that maturity assessment is essentially backward-looking. It considers how far a technology has progressed, which is useful information in judging whether enough effort has been expended to reduce the risk of future problems. However, this says nothing about the future – as the financial world says in its standard disclaimer, 'past performance is no guarantee of future results'. Hence, some caution needs to be exercised in considering a new technology to be problem-free simply because it has reached a certain maturity.

The assessment process should reduce the risk of future problems but it will not eliminate them. For example, it is not unusual for a technology that has worked well in one application or environment to be problematic in another where the duty cycle or operating environment is different. The engineering development process trials the new technology in an increasingly realistic environment, as progress is made through the TRLs, but it is not until the real operating environment is reached using production material that victory can be declared.

One way of addressing this issue is, at the same time as assessing current technology maturity, to undertake an expert review of future problems and risks, methods of identifying them, and measures that can then be taken to overcome them. Again, the aim is to reduce risk whilst acknowledging that risk can never be totally avoided. An approach to this forward-looking assessment has been codified by under the title 'Advancement Degree of Difficulty' or AD^2. Chapter 7 looks in some detail at risk identification and avoidance.

3.15 Concluding Points

The principle that new technologies pass through identifiable phases of development is an important one to grasp when considering how to manage and plan the development of new technologies. There is an element of universal truth about the concept – rather like human development from cradle to grave, there are elements one would prefer to avoid. This, of course, is not possible, and all the examples from engineering history show that short-cuts cause problems – the process can certainly be made more efficient, but all phases of maturity have to be worked through. The TRL and MRL scales give a common language for this process and, in particular, can be used as a means of communication with nonexpert parties such as general managers or investors.

References

The first five references provide background to the original development of the TRL system within the US space industry:

3.1 NASA Headquarters Oral History Project, Edited Oral History Transcript, Donna L. Shirley, Interviewed by Carol Butler, Norman, Oklahoma – 17 July 2001, https://www.jsc.nasa.gov/history/oral_histories/NASA_HQ/Herstory/ShirleyDL/ShirleyDL_7-17-01.htm
3.2 Sadin, S.R., Povinelli, F.P., and Rosen, R. (1988). NASA technology push towards future space mission systems. *Acta Astronautica* 20: 73–77.
3.3 Mankins, J.C. (1995). *Technology Readiness Levels: A White Paper*. NASA, Office of Space Access and Technology, Advanced Concepts Office.
3.4 Software Technology Readiness for the Smart Grid — Cristina Tugurlan, Harold Kirkham, David Chassin Pacific Northwest National Laboratory Advanced Power and Energy Systems Richland, WA, PNSGC Proceedings, 2011.

3.5 Nolte, W.L. and Bilbro, J.W. (2008). *Did I Ever Tell You About the Whale?: Or Measuring Technology Maturity.* Information Age Pub Inc.

The UK automotive industry has developed a system for its specific needs:

3.6 Automotive Technology and Manufacturing Readiness Levels: A guide to recognised stages of development within the Automotive Industry, Automotive Council UK, January 2011

The only formal international standard is:

3.7 ISO 16290:2013 Space Systems – Definition of the Technology Readiness Levels (TRLs) and Their Criteria of Assessment

A spreadsheet method of TRL estimation is provided in this reference:

3.8 Technology Readiness Level Calculator, Air Force Research Laboratory

This document provides an approach for using TRL estimation in the context of early-stage investment:

3.9 Measuring Technology Readiness for Investment – The Manufacturing Technology Centre & Heriot-Watt University, March 2017

4

Aligning Technology Development with Business and Manufacturing Strategy

4.1 Introduction

The development of a new product or a new technology, complex, interesting, and demanding though it may be in its own way, is really only as useful as the manufactured product that emerges from it. Without that useful output, sold either to the public or to a business customer, technology development is merely an intellectual exercise. It therefore makes sense that the development process, from the outset, should be aligned as closely as possible with the business and manufacturing strategy for exploiting it commercially. The emphasis here is on the word *outset*; seeking alignment after much effort has been invested is always less efficient and will run into 'not invented here' issues.

At the tactical level, this could simply mean that a new product should be designed so that it can be made easily – 'design for manufacture' – and that is always good practice.

However, there are some more fundamental considerations, such as:

- Who will make the various components of the product and who will assemble it?
- Will there be just one product, will there be various options, or will there be some level of customisation for a specific purchaser?
- Will it be simply sold and forgotten (unlikely), or will there be some supporting services afterwards?
- Will the manufacturer be selling the product as a service, effectively taking responsibility over the product's whole life?
- Will the manufacturer have to take some responsibility for disposing of the product at the end of its useful life?

4.2 Business Context

Manufacturing is often described, rather simplistically, as the conversion of raw materials into finished products. Whilst this is an accurate description of

Managing Technology and Product Development Programmes: A Framework for Success,
First Edition. Peter Flinn.
© 2019 John Wiley & Sons Ltd. Published 2019 by John Wiley & Sons Ltd.

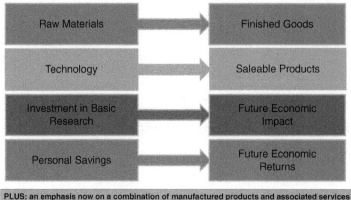

Figure 4.1 Perspectives on manufacturing.

the mechanics of manufacturing, the process can be considered much more broadly: as well as being a material conversion process, it is also the route by which new technologies are converted into useful, saleable products; linked to this, it is the means by which more basic research is converted to economic output (something which always interests governments); and it also a means by which money such as personal savings can be turned into future economic returns, rather than just being kept under the mattress. These points are shown in (Figure 4.1):

So, in summary, rather than simply being the conversion of raw materials into finished goods, manufacturing plays a more strategic economic role in a number of different ways. The value-adding activities also do not stop

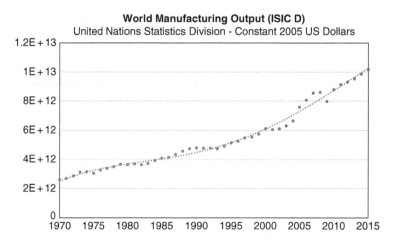

Figure 4.2 Global manufacturing output, UN Statistics Division.

at manufacturing; logistics, delivery, installation, and service all make an economic contribution and widen the definition of 'manufacturing'.

For these reasons, manufacturing is big business. It represents about 15% of the world's economic activity with a global turnover of some $10 trillion at 2005 prices (Figure 4.2) – see Ref. 4.1.

It is also a growing activity: over the period 1970–2015, the world's output of manufactured goods more than trebled, representing an annual growth rate of just under 3%.

4.3   Basis of Competition

It also follows that manufacturing is a very competitive, international activity. Manufactured products are mobile and can be shipped around the world quickly and cheaply. Any new product or technology therefore must be capable of fighting its corner against direct competition, as well as against alternatives or substitutes.

In this situation, the developers of new products and technology must decide on what basis they wish to compete. Michael Porter's work from the 1980s – Ref. 4.2 – gives some useful guidance. He identified three broad strategies that companies could consider (Figure 4.3):

1) Cost leadership
2) Differentiation
3) Focus

Taking each strategy in turn, *cost leadership* is the most direct form of competition and relies, as the name implies, on cost management being the driving force of the firm. This typically implies high market shares, economies of scale, heavy investment, tight cost control, and design for low-cost and possibly local manufacturing, and easy service. In cost or price terms, this approach sets the benchmark or reference point for other forms of competition.

Differentiation, on the other hand, places more emphasis on design, brand image, durability, and high resale value. The product will command a higher price, although volumes and market share will be lower. The higher price will have to be in balance with the benefits of the product, and there will be

	Source of Competitive Advantage	
	Cost	**Differentiation**
Market Scope — Narrow		*3. Focus*
Market Scope — Broad	*1. Cost Leadership*	*2. Differentiation*

Figure 4.3 Sources of competitive advantage.

a constant battle with lower price/lower specification products who will eat away at the differentiator's market share.

The *focus* strategy is essentially a niche approach where the appeal is to a narrow range of closely defined customers with whom an almost personal relationship can be built. Direct competition is relatively weak in this situation, but substitutes or alternatives must always be borne in mind.

For a firm developing a new technology or product, the appropriate strategy needs to be borne in mind to ensure that the new offering, and its manufacturing approach, are consistent with the chosen path. If the development is taking place within an established company, the approach will already be set. Where the company is a new entrant, realistically, the 'niche', or at best the 'differentiation', approaches are likely to apply on the basis that the costs and investments associated with 'cost leadership' will be too much for a new firm.

It could be argued that a fourth model of competition has emerged, or is emerging – concerned with the business model defining the relationship between supplier and customer. To a large extent enabled by technology, in the engineering world it revolves around ownership of the productive asset and the selling of it as a service by the manufacturer, rather than selling it as a product to an operator. In some fields, such as those created by Uber and Airbnb, the new business model has clearly created a new basis of competition.

4.4   The Value Proposition

Clarity over the positioning of a new development in the marketplace is clearly a critical issue, brought into focus by research findings that suggest that some 40% of new, launched products fail to be successful – Ref. 4.3. The failure rate for initial ideas will be much higher as most ideas fail to make it even to the first hurdle.

An effective way of addressing how a product should be positioned in a market is through a *value proposition*. The concept of the value proposition can be traced to work by the Atlanta office of the McKinsey organisation. Michael Lanning and Edward Michaels of that office wrote a McKinsey staff paper in June 1988 – Ref. 4.4 – which spoke of the value proposition as: 'a clear, simple statement of the benefits, both tangible and intangible, that the company will provide'.

They went on to say: 'And we are not talking about vague benefits, such as "good quality." We mean concrete, observable features of the product or service'.

For example, the Airbus A380 is positioned as offering: 'a 15 per cent lower cost per seat-kilometre than its rival large airliner' – quite a clear statement of intent, although of course a large and complex product such as an airliner will have a long list of features and benefits which will appeal to the purchaser, the operator, and the flying customer but this is one clear, headline.

A good way of working on a value proposition, particularly if conducted as a facilitated group activity, is through a *value proposition canvas*, of which there are numerous, and sometimes copyrighted, models. They all follow a similar pattern in the form of a matrix relating the benefits of the product to the needs of the customer, ideally on an item-by-item basis.

Within the workshop environment, the customer's needs are documented and the benefits or features provided by the supplier are listed. Work is then done to Identify positively where benefits meet the customer's specific needs or where they avoid or reduce a specific customer problem. Each point should be specific and quantitative wherever possible and broad generalities should be avoided, as they are difficult to translate into engineering form.

Positive benefits are usually easy to identify – reduced operating cost per seat-kilometre in the Airbus example above. Problems avoided, or 'pains relieved', require more mental agility but much improved reliability, for example, or reduction in service problems, might be examples of this. Again, these points can be quantified.

Doing this analysis as a group will open up new insights and put a different perspective on what is being developed for the customer.

A (hypothetical) example for a simple, low-cost robotic system is shown in Figure 4.4.

The most important element of this work is the headline statement, which acts a guide and focal point for the new technology or product. This statement can also be built on by adding in more detailed product definition material and further information about how customer needs will be met – see Chapter 5

VALUE PROPOSITION MAP – *Low cost, safe, easy-to-program robotics solution*			
SUPPLIER		CUSTOMER	
BENEFITS	Beneficial features	Positive needs	NEEDS
Low cost *2 kg payload capability* *700 mm reach* *6 axes* *± 1.0 mm positional accuracy* *Intrinsically safe* *Capable of human/robot cooperation*	*Cost <£5k* *No special power supply required*	*Low cost to enable payback at low utilisation rates* *Easy to install and operate* *1.5 – 2.0 kg payload*	*Higher pay back automation solutions* *Modest payload for simple tasks* *Operating within 2 m × 2 m envelope*
	Avoiding features	Problems to be avoided	*mm accuracy*
	Safety case developed *Built-in "stop" if collides* *Intuitive programming system*	*Robot has to be operated within safety enclosure* *Safety case has to be developed* *Specialised training needed for programming*	*Capable of operating in open workshop* *Easy to re-programme*

Figure 4.4 Example of a value proposition map.

for more details. From this work, it will become more evident how the new product will compete – whether it could be a niche product, a cost leader, or a differentiated offering – and whether this is consistent with the company's chosen strategy.

4.5 Industry Structure

A further shaping factor is the structure of the industry into which the new product will be sold. The product may simply be made by one company and sold through its own sales outlets or through dealers, wholesalers, or retailers. Alternatively, the product may be incorporated into a further, more complex end-product, such as an aircraft or road vehicle. Consider, for example, how a tail-lift for a truck body, not a particularly complex product in its own right, finds its way into the market (Figure 4.5).

Identifying the real customer or customers, and other involved parties, is difficult in this situation but it is essential to understand who is, or are, the real customers, as well as understanding the operating environment for the product and who understands that particular point. These points underpin effective marketing.

The structure of the market may also determine whether one standard product is sufficient, or whether a range of options is needed, and how the product

STAGE IN PROVIDING THE FINISHED VEHICLE	COMPANY INVOLVED
Customer of logistics company	Retail chain with exacting delivery schedules
Vehicle operation	Operated by logistics company with contract to retailer
Operational, finished vehicle	Owned by truck leasing company
Vehicle provider	Chassis-cab provided by a major, international truck manufacturer Vehicle completed by a regional body-builder
	Tail-lift sold to body-builder and incorporated into vehicle body, then connected to a power-source on the chassis-cab
Truck manufacturer	Designs & assembles finished chassis-cab
Suppliers to truck manufacturer	Several tiers of suppliers to chassis-cab builder and to body-builder

Figure 4.5 Example of complex industry structure.

is ordered, as well as the lead time. These factors all have a bearing on new product development activities and need to be recognised explicitly during the commercialisation phase of product development.

4.6 Routes to Commercialisation

Having established the broad approach to defining a competitive product and the product's positioning in the marketplace, the route to commercialising the new product needs to be determined. It is not the purpose of this book to discuss the advantages and disadvantages of different routes. However, the choice of route is relevant to the technology development process in the sense that the design of the product carrying the new technology could be affected by how it is manufactured and how it is sold.

There are many routes to commercialisation, and the chosen approach will depend on the market and industry structure, as well as the barriers to entry, the status of the company developing the product and the funds available. Circumstance make dictate only one feasible route, especially if funds are limited (Figure 4.6).

In an established business, the route to market for a new technology or product is likely to be through existing, or adapted, sales, logistics, service, and support channels.

For a new undertaking, there are four possibilities:

1) Set up own manufacturing, sales, and support channels – which could require substantial time and investment but may be feasible if it is a fundamentally new market and a start can be made at modest volumes.
2) Subcontract manufacture and franchise sales and support – a similar strategy to the above but working with partners who may share some of the costs and liabilities (and rewards).
3) Sell the technology to an established company – which may be the only way forward if there are substantial barriers to entry to the planned market due, for example, to the manufacturing volumes required or the regulatory requirements of the market.
4) License the technology – which retains control of the product but which passes most of the reward to other parties and where copying is always a possibility.

Combinations of these strategies are also possible. For example, the company could retain some manufacturing capacity, from which it can learn and develop, but also license additional capacity in other territories or perhaps licence an earlier generation of products to protect the company's know-how, provided this does not then create a new competitor.

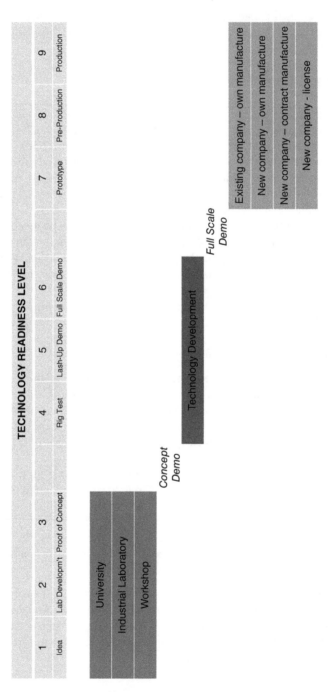

Figure 4.6 Alternative routes to commercialisation.

4.7   Satisfying a Range of Customers

Henry Ford famously remarked, of the Model T in 1909, '*Any customer can have a car painted any color that he wants so long as it is black*'. He went on to say, '*it will be so low in price that no man making a good salary will be unable to own one*'. Black paint, incidentally, dried more quickly and was therefore cheaper.

He was clearly pursuing a classic cost leadership strategy and using absolute standardisation as one means of achieving this. His strategy worked well for quite some time, but he eventually had to bow to competition and be more flexible. In the twenty-first century, the expectation is that customers can have almost anything they want.

When developing new products, a clear strategy is needed in terms of deciding how the requirements of different customers will be met. Ultimately, this is a business decision, and more variety generally means more cost but more volume. The choice has a major impact on manufacturing, purchasing, and logistics strategy and, in turn, on how products are engineered.

In principle, there are many options:

1) Just one standard product
2) A range of standard products
3) Standard products with pre-engineered options
4) Standard products but with customer-specific changes
5) Special product built to customer's requirements

In practice, different products and different industries have different approaches. Consumer goods tend towards the upper end of this list whilst products or system for business or industry tend to be more heavily customised. The general trend is towards *mass customisation* – a very wide variety of choice but at mass manufacturing costs.

The automotive industry is particularly adept at satisfying a wide range of customer needs with apparently different products whilst actually having a modest range of common platforms. Volkswagen Group, for example, manages at the time of writing to create almost 20 different models from its one Golf platform, covering VW, Audi, SEAT, and Skoda products.

Achieving this level of diversity from a common set of parts requires a planned and strategic approach from the outset – and it has taken Volkswagen many years and several iterations to arrive at this position. In principle, it means that the wide range of possible options need to be taken into account at the early stages of design. Split lines or a modular approach need to be built in, but in a way that does not create too much additional cost. In fact, there is always a trade-off to be made between building in flexibility, on the one hand, and minimising product cost on the other hand.

The references above to small car platforms have, of course, built on the experience of several decades of producing a range of products from common platforms. When a substantially new product is being developed in a new or emerging company, it is much more difficult to predict what options customers might want, and it might take several generations of product to reach a situation where different customer's needs are understood and can be met by an efficiently engineered family of products.

This approach also acts as a strong driving force when it comes to manufacturing strategy.

4.8   Linking to Manufacturing Strategy

In an established company, there will almost certainly be a manufacturing strategy, possibly an unspoken one, that will shape the development of new technologies and products. It will cover topics such as:

- What the company chooses to make itself versus what it buys externally – 'make-vs.-buy'
- How far afield it looks for its suppliers – close to the company or further afield with attendant transport costs and time lags
- Where it chooses to locate its manufacturing, which could be one central location or could be distributed globally
- What manufacturing processes it will consider or has the facilities and skills to pursue
- How it accepts orders and promises delivery timescales
- How it deals with the needs of different customers
- How it aligns its policies with government pressures

The development of new technologies and products cannot proceed very far without taking these points into account. For example, if a company has particular expertise in, and facilities for, certain manufacturing processes, it will want to use them in a new product, which, in turn, means that the product will have to be designed around them. In the case of an early-stage company, some idea of likely manufacturing processes will be needed to assure manufacturing feasibility and to give some realistic idea of costs. In both cases, work will be needed with suppliers for purchased components for similar reasons.

It can be seen that it makes sense for technology and product development to proceed in close harmony with manufacturing and commercial strategy, which leads to the question – how best to achieve this?

4.9   Core Principles of Managing the Interface

The thrust of the preceding sections is to point out that technology development cannot proceed very far in isolation from broader business factors. In an established business, this means different groups collaborating together to bring in their combined experience and specialist knowledge. In a new business, it means investing effort in business considerations, however that is done, from an early stage. If these points are ignored, it is unlikely that an optimal solution will emerge, so there will be a need for backtracking or rework to bring the technology development in line with other factors.

The following principles should be borne in mind:

1) These considerations and activities should be programmed in from the outset.
2) Manufacturing and commercial objectives should be set from an early stage, as well as product objectives.
3) Engineers should adopt the principle of early release of preliminary information, accepting that it is imperfect, rather than waiting until all the details have been completed.
4) There should be intensive communication between all the interested parties from the earliest possible stage.
5) Structured methodologies can be used, typically in a collaborative environment, to work through these points.

The organisation's culture will have a considerable bearing on the success of this work. It is reliant on a collaborative, blame-free culture to manage the interfaces and interactions between designers/developers, and commercial and manufacturing experts. These are often groups with different life philosophies, timescales, levels of pragmatism, and commercialism These points often underlie tensions between such groups but, in reality, one can't exist without the other. Ideally, the driving philosophy should be that mentioned in Chapter 1 – an engineer can do for a dollar what any fool can do for two.

The symptoms of poor interfaces are easy to spot – late redesigns, lack of empathy, product cost targets not met, and quality problems in manufacturing due to difficult operations – are just some examples. However, these issues can be overcome, which makes a big difference.

4.10   Design for Manufacture Methodologies

The principles established above lead into the design of individual components or assemblies.

'Design for manufacture' is a well-established discipline at the level of the individual component, or the assembly or sub-assembly operations. Usually, it is termed 'design for manufacture and assembly' (DFMA). Related to this there is 'design for life-cycle operation', 'design for maintenance', and 'design for disposal', leading on to the more general umbrella term 'DfX'.

There are many established methods for optimising the design in relation to these factors and many useful supporting publications. The classic reference in this field, Ref. 4.5, is by Boothroyd, Dewhurst & Knight, running over 700 pages and covering both assembly and most of the common component manufacturing processes. There are other books covering the same ground. Many consultancies operate specifically in this field, and there are technical conferences devoted just to this subject. Additionally, there are hundreds if not thousands of booklets from industries or individual companies describing how best to design individual components in a particular material or with a particular manufacturing process in mind.

At the level of the component, the basic guidelines relate to:

- Reducing or simplifying the features or complexity of a component
- Using readily available materials
- Not imposing difficult-to-achieve tolerances and surface finishes
- Using standardised sizes, e.g. for drilled holes
- Working to the limitations of the specific manufacturing process, e.g. spring-back, draft angles, die withdrawal angles, or bend radii

Similarly, there are straightforward generic guidelines for the development of assembly operations:

- Reducing the number of parts or integrating multiple parts into one part (which runs somewhat counter to the component-level guidance)
- Analysing the real need for each individual component
- Use of standard, low-cost parts
- Use of click fasteners rather than threaded nuts/bolts/washers (which are also often the source of problems in service)
- Providing easy access, location, and vision for assembly work
- Avoiding very small, fiddly parts or conversely heavy and cumbersome parts
- Presenting parts in a way that eases assembly
- Reducing operations that require special tools or fixtures
- Assembling from one direction, using gravity, providing self-locating or self-aligning features, reducing assembly forces
- Reducing the need for orientation of parts through symmetry
- Reducing post-assembly adjustment or setting operations

Organised or tabular worksheet methods of functional analysis can also be applied in a group workshop environment to optimise designs, sometimes supported by scoring systems that rate 'design efficiency' or ease of assembly or manufacture.

Putting aside specific aspects of a particular component or assembly, the key point is that DFMA activities should be built into the design process from the outset and not conducted as an afterthought or follow-on process. Estimates suggest that some 70–80% of manufacturing costs are established by the design activity, so it pays to put effort into DFMA at an early stage. Once a design has been established, and probably tested and analysed, it becomes increasingly difficult and costly to change it and undo the investment already made in it even if the manufacturing cost reasons are quite compelling.

4.11   Design for New Methods and Materials

The points above relate primarily to existing material and manufacturing methods where there are already well-established experience and good practice. The aim of design for manufacture work in this situation becomes one of optimisation and avoidance of poor solutions.

When a new material or a new method of manufacturing become available, the natural tendency is to design parts that closely resemble those which went previously using the earlier methods and materials. For example, the first plastic automotive body components looked remarkably similar to their steel predecessors. This is a product of conservatism and a lack of experience of what does, or does not, work in practice, plus the fact that replacing a conventional part on a one-for-one basis puts further boundaries on what can be achieved.

Once experience is gained, design concepts change more radically to take advantage of the new situation, not just at the level of the individual component but also in terms of the surrounding parts, the roles they play, how they are attached, and how they are configured.

An example is shown in Figure 4.7 of how a small component might be made differently, and radically so, using additive manufacturing methods. The new part might also be extended to perform some of the functions of the parts to which it attaches, resulting in one consolidated part in place of several components.

All these factors should be considered at an early stage and will ideally form part of a technology development programme preceding a product development programme. Leaving until later will reduce the opportunity to take full advantage of new technologies.

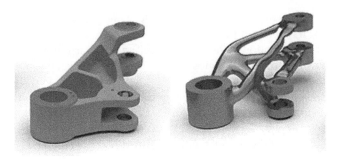

Figure 4.7 Component manufactured by additive methods compared with conventional.

4.12   Design for Connectivity – Internet of Things

Some form of internet connectivity back either to the manufacturer or the asset owner is becoming a mandatory requirement for a wide range of engineering products from simple household electrical appliances through to major pieces of capital equipment. Computer diagnostic systems have done this since the 1980s. The purpose of doing so varies and could include:

- Simple remote control of devices by the owner
- Online monitoring of performance and subsequent adjustment of, for example, a production process
- Software maintenance or updates
- Measurement of service environment ahead of maintenance
- Measurement of duty cycle to aid future designs
- Anticipation of failure by, for example, measuring vibration, noise, currents, and voltages
- Measurement of usage where the product is sold as a service

At a practical level, the sensors and methods of communication need to be built into a product at an early stage to facilitate these features. Linked to this, the form of data analysis for information from the operating product need to be designed alongside the product itself. This, in turn, requires the nature of the service offered, and its associated business model, to be written into the business plan for the company as a whole. Hence, the use of IoT goes well beyond a purely technical issue.

4.13   Design for Environmental Considerations

Environmental factors play a big part in the development of new technologies and products. In fact, enhanced environmental performance, linked to energy generation or consumption, is one of the major drivers of new technologies. These factors can affect product development in several ways:

- Choice of materials from which the product is made, avoiding those which are in short supply or have environmental sensitivities.
- Reducing the energy consumption of the product itself and the harmful emissions from it. In many industries if not most, there are legal requirements to be met in relation to emissions.
- Reducing energy consumption of manufacturing processes, use of harmful chemicals and emissions or discharges from these processes. Again, legal requirements probably apply.
- Minimising the packaging of the product and using readily recycled materials.
- Planning the disposal or reuse of the product by, for example, choice of readily recycled material or simplifying disassembly.

A full life-cycle cost of ownership assessment is one way of taking a holistic view of these factors and could be used to evaluate alternatives. These points can all be built into the plan for programmes of technology or product development and managed as requirements of that plan, alongside all other requirements.

4.14 Concluding Points

The key point of this chapter is the fact that the creation of new technologies or products cannot proceed as an independent activity, isolated from the rest of the world. Its ultimate aim is to create a successful business. This means that these broader considerations should be taken into account from the earliest possible stage. Factors to be considered include:

- How will the product compete?
- What are the key features that will appeal to a buyer?
- How is the market structured?
- Who are the customers?
- How will it be commercialised?
- How will the needs of different customers be met?
- How will the product be priced, and what are the cost targets?
- What environmental performance is required?
- What will the manufacturing strategy be?
- How will it be supported in service, and will there be a real-time connection?
- How will the design detail be developed so it can be manufactured easily?

This is quite an array of questions, and they cannot all be answered on Day 1. Equally, *failing to consider them* from an early stage will almost certainly result in the wrong product or a difficult process of redesign where earlier work has to be repeated.

This highlights one of the most fundamental dilemmas of technology and product development, to which there is no easy answer. The most successful companies have found ways in which different areas of the company are able to cooperate in bringing the widest possible perspective to early stage work. It requires a very positive culture for it to work. All involved to be able to take a strategic view; for example, there needs to be a clear view of manufacturing and supply strategy. The engineers originating the ideas need to be especially tolerant or thick-skinned as their ideas are put through the mangle.

Senior managers have their role to play in creating the right climate. They also need to be supportive but willing to take difficult decisions, killing off initiatives that are going nowhere. It might be thought that these comments are solely the province of the larger company. Small companies face exactly the same issues. They need to go through the same thought processes and make the same decisions. In their favour, this may only involve a few people and limited bureaucracy or levels of authority.

References

This comprehensive international database can be used to analyse GDP and the contribution of manufacturing (and other sectors) country by country:

4.1 United Nations Statistics Division — National Accounts Main Aggregates Database, GDP and its breakdown at constant 2005 prices in US Dollars Michael Porter's book is one of the classics of business management:
4.2 Porter, M.E. (2012). *The Competitive Strategy: Techniques for Analyzing Industries and Competitors*. New York: Free Press.

This paper provides some interesting information about how frequently new products succeed or fail in the marketplace:

4.3 Castellion, G. and Markham, S.K. (2013). Perspective: new product failure rates: influence of argumentum ad populum and self-interest. *Journal of Product Innovation & Management* 30: 976–979.

This McKinsey paper is the origin of the concept of the value proposition:

4.4 A business is a value delivery system Michael J. Lanning and Edward G. Michaels, McKinsey staff paper dated June 1988

Boothroyd's book is another classic, this time dealing with design for manufacture and assembly:

4.5 Boothroyd, G., Dewhurst, P., and Knight, W.A. (2011). *Product Design for Manufacture and Assembly*, 3e, 650. Boca Raton, FL: CRC Press.

5

Planning and Managing the Work

5.1 Introduction

This chapter is concerned with how development programmes, of all types, should be managed to achieve the desired results. In this respect, it deals with 'project management'. The context is programmes for new technologies or new products, and the term 'project' is used in this chapter to refer to these programmes of work.

At one level, the development of new technology and products can be seen as an essentially technical process, concerned with, for example, ideas, drawings, calculations, and tests. The core processes of development do obviously centre on activities of this type and many more. However, that is not to say that 'management' has no part to play, in either a facilitating or overseeing role. In particular, the effectiveness of development will only be as good as its planning and subsequent direction. This is increasingly the case as technologies or products advance in development maturity. Inevitably, the projects associated with them become bigger and more complex as they move towards production, and hence, they become potentially more difficult to manage. Effectiveness will also depend on how the people staffing the project are led and how the host organisation(s) function.

At one end of the spectrum, there are small-scale research projects involving just a few people over a period of months or a few years. At the other end, there are major projects, such as the design and construction of aircraft carriers where the engineering effort is measured in millions of man-hours and which may run over 5–10 years.

Clearly, large projects require more management than small ones, but some basic principles apply to every situation. A distinction also needs to be drawn between projects with a high level of built-in uncertainty versus those where the tasks making up the project are more concrete.

Every project contains an element of learning, but technology development projects, with a significant level of uncertainty, are as much about learning as they are about the end-result. Some projects are deliberately set up in

Managing Technology and Product Development Programmes: A Framework for Success,
First Edition. Peter Flinn.
© 2019 John Wiley & Sons Ltd. Published 2019 by John Wiley & Sons Ltd.

this way, for example, the US 'X' aircraft projects (X-1, X-15) in the 1950s. Major 'delivery' projects, on the other hand, focus on very specific outputs that are usually contractually specified in terms of time, performance, quality, and cost.

In this context, research, technology, and software projects may have defined aims, but achievement of those aims involves a lot of exploration, iteration, and false starts. Projects of this type cannot be neatly programmed and managed as a series of predictable tasks. That is not to say, however, that they will not benefit from some basic management attention, which will increase the probability of project success.

Large-scale product development projects, on the other hand, can be programmed in much more detail, and typically such projects are subdivided into multiple tasks, or subprojects, involving many people over an extended period. Classic project management methods apply in these situations, and such projects will employ professional project managers or project management organisations to handle them. That is not to say that problems will not still arise, but, if identified clearly and early, they can usually be contained.

This chapter, and this book more generally, places the emphasis on work of a modest scale but with relatively high levels of uncertainty. The principles, however, are valid for all projects.

5.2 The Basics

It is surprising how many projects are attempted without paying attention to even the most rudimentary planning. This leads to a waste of good effort. Attention to just a small number of basic points of good practice, which apply to all projects whatever their size or level of uncertainty, can make quite a difference. These points can be summarised as follows:

1) *Develop a project mandate.* A short, one-page summary provides the background, purpose, and objectives for the project.
2) *Write a project description.* A narrative account of the details of the project – the process of writing this will be as valuable as the end product.
3) *Compile a milestone plan.* A summary, in chart form, of the project's overall timescale and its intermediate checkpoints.
4) *Estimate a project budget.* A calculation of how much money is needed and when; it also includes an estimate of the manpower required.
5) *Draw up a risk analysis.* A chart showing the areas where people have expressed concerns that may affect the project outcome.
6) *Draw up a responsibility chart.* A list of the people required, ideally by name, and their roles, responsibilities, or specialisations. This is sometimes known as a 'RACI' chart (responsible, accountable, consulted, informed).

Attention to these points and their wide communication will get a project off to a good start but does not provide a complete solution. Each point is described in more detail below, but, first, two different approaches to project management need to be outlined.

5.3   Different Approaches

The classic approach to managing work of the type described above is to break it down into tasks or activities, build up a cost, timescale, and resource for each task, and then reconstruct it in the form of an overall programme, usually presented in the form of a network. This approach is fine where the tasks are reasonably stable and predictable, do not interact to a high degree, and do not depend on the outcome of earlier tasks, other than their completion. It can be scaled up to very large projects and lends itself to digital methods of project control. Where the project involves learning and hence uncertainty, as is the case with most technology development projects, trying to plan ahead in detail is a futile activity – most of the tasks will have changed by the time they are reached.

An alternative approach is to break the project down into a series of intermediate goals or milestones, representing 'states' through which it must pass. This approach is sometimes known as 'goal-directed project management' and is described in detail in Ref. 5.1. Each goal takes the form of a statement such as: 'when …… .has been achieved'. Detailed activity or task planning is restricted to the upcoming goals, whilst those later in the project are not planned until they become relevant and the learning from earlier stages has been absorbed.

5.4   Different Forms of Project

Figure 5.1 illustrates the different forms that projects can take. The danger zone is the top right-hand corner where projects are both complex and involve a high degree of learning or uncertainty – these usually go badly. Projects can also drift by accident into this zone; for instance, a major product development project, apparently of low uncertainty, can find itself in the danger zone if elements of its technology prove to be underdeveloped or if the scope of the project creeps beyond what was agreed.

Technology and product projects can start using the goal-directed approach and then move into the classic approach for the commercialisation phase. The key to this transition is knowing when the technology development has reached a point where its later risks are relatively low and hence the classic approach can be used. Chapter 3 covers the topic of technology maturity.

DANGER ZONE!

High Uncertainty	Small research projects	Large research projects Public-sector IT projects

Drift!!

Low Uncertainty	Routine work	Major construction projects Major product development projects
	Low Complexity	**High Complexity**

Figure 5.1 Different forms of project.

PROJECT NAME – xxxxx
PROJECT SPONSOR – AB Jones, managing director
PROJECT LEADER – CD Smith, research centre manager
BACKGROUND - yyyyyyy.
PURPOSE - The purpose of the project is to zzzzz.
SCOPE - the proposal is to xxxxxx.
AREAS OF WORK NOT INCLUDED - yyyyy is not part of this project.
GOALS • aaa • bbb • ccc • ddd
CONSTRAINTS & GUIDELINES – the project must be completed within the current year, i.e. by 31st December 2017
PROJECT BUDGET - £500,000 including expenses but excluding VAT
SUCCESS CRITERIA – acceptance of the system by the stakeholders and agreement to proceed with a subsequent implementation phase

Figure 5.2 Example project charter.

5.5   The Project Mandate or Charter

The project mandate or project charter is the first step in formally defining a project. An example is shown on Figure 5.2. It should be confined to one page, to make it manageable, and used to gain acceptance in principle for a project by laying out the basic approach – what it's called, who is supporting it, who is running it, what are its aims, goals, and objectives, what it will cost, and when it will be complete. If these points can't be agreed, there is no point in going further. As with most early-stage activities, the value comes more from the dialogue that the mandate creates rather than the finished document.

5.6   Project Description

Once the mandate has been agreed, the project can be fleshed out in more detail. Figure 5.3 provides a summary of the potential content of a project description. The details will vary from one project to another, but this is a starting point. The volume of content will also vary widely: a small project might only need four or five pages. Large, complex project may need hundreds. As

1	Introduction	Some background to the project, its origins and its broad aims
2	Objective	A succinct statement of the aims of the project
3	Scope	The boundaries of the project, what is included, and what is excluded
4	Business proposition	The basis on which the product and/or service will be marketed to customers as an attractive value proposition
5	Technological approach	A summary of the technology that will form the basis of the project, both known technologies and ones that are relatively new
6	Participants	A listing of the main contributing organisations or departments/groups
7	Aims of the participants	The underlying aims of the participants and what is driving them
8	Finances	A breakdown of the cost of the project
9	Programme of work	A description of the principal work elements of the project
10	Programme management	The approach that will be taken to oversee the project
11	Milestone plan	An overall timing chart or milestone plan for the project including its completion date and the principal way-points during the project
12	Risks	A summary of the project's main risks and the approach to managing them
13	Relationships to other projects	Links to other current, past, or future projects
14	Health & safety	Any EH&S risks that go beyond the organisation's normal profile and that may therefore need special measures
15	Intellectual property	How any IP arising from the project will be documented, owned, and protected
16	Communications	How the work of the project will be communicated before, during, and after its execution

Figure 5.3 Example project description.

with the mandate, the value comes more from the dialogue that the document creates rather than the finished item, although in the case of the project description it is likely to prove useful as a future point of reference and it should be updated as the project proceeds.

This structure might be considered the minimum for a reasonably comprehensive project description. Others would advocate more detail. For example, in his book *Total Design*, Stuart Pugh (Ref. 5.2) recommends over 30 categories of information. There is always a balance to be struck between too little and too much information: too little and the product might not meet market or customer requirements; too much and the scope for initiative is reduced. As an example of the latter, the author has experience of a publicly procured rail project where the buffet car's microwave oven enjoyed five pages of specification. In this situation, there is very little opportunity for innovative solutions and problems are likely to be blamed on the specification.

5.7   Timing Charts

A timing chart or Gantt chart represents the simplest and clearest way of laying out a basic plan. The method was developed originally in the early 1900s by Henry Laurence Gantt, who could trace his ancestry back to the Norman invasion of England in 1066. Interestingly, William the Conqueror's success has been put down to careful planning and not launching his 'project' until everything was in place.

Gantt himself was a mechanical engineer and a colleague of F.W. Taylor, with whom he developed the principles of 'scientific management'. He is still remembered by the ASME Henry Laurence Gantt Medal.

It is essentially a graphical method of planning the work on a project, showing interdependencies, and recording progress. It was first applied as a production work scheduling tool in a factory environment – see 1903 paper 'A Graphical Daily Balance in Manufacture'. It does rely on the activities being quite finite and predictable, with clear start and finish points, but even where this is not the case, a timing chart provides a basic frame of reference from which to manage the work. Figure 5.4 is an example of a very simple chart for a hypothetical technology development project.

The approach can be expanded out into very complex networks for large-scale projects. The sequence of activities can be shown in terms of which tasks must precede others or finish concurrently. Critical paths, which determine the overall timescale of a project, can be identified and subsequently managed. Activities can be assigned to resources – either people or physical assets or computing resources – and hence the loading of resources can be estimated. This enables detailed cost estimates to be compiled as a function of time. Actual cost expenditure can be monitored and compared with the value of the work completed – sometimes known as the earned value.

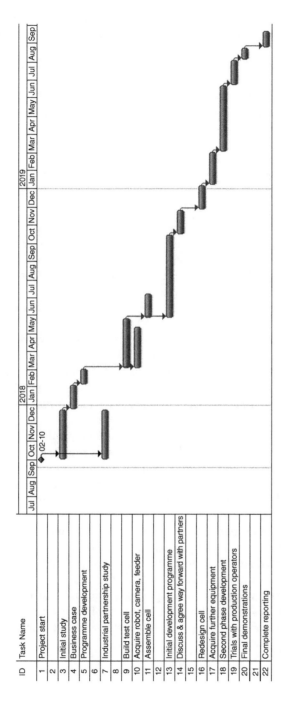

Figure 5.4 Example timing plan or Gantt chart.

Detailed plans of this type are relevant to high complexity/low uncertainty projects such as construction programmes where the details can be predicted accurately and well in advance. Several large-scale software programmes are available for such projects. Using such tools, the concentration of management effort is on identifying problem tasks, which are off-programme, and then remedying them. If too many tasks become problematic, then such projects can become close to unmanageable, emphasising again the importance of reducing risk before projects are initiated.

5.8 Milestone Charts

An alternative approach, especially valid for early-stage technology development projects with more in-built uncertainty, is a milestone chart, of which Figure 5.5 is an example. The project is first broken down into a small number of separate result paths, which represent the principal streams of activity in the project. Milestones are then agreed, representing key points of development in the project – there are 12 in the example.

These milestones need to be spread evenly across the project, acting in a similar manner to 'way points' in aircraft or sea vessel navigation. The milestones should have clear deliverables associated with them, and it should be easy to judge whether a milestone has been achieved.

The process of agreeing to these milestones will create a common understanding about the details of the project if it is carried out as a collaborative, team activity. Conversely, an imposed plan will have little ownership and will not be respected.

The planning method outlined above concentrates on results – 'what' must be completed – and provides a clear basis for monitoring whether the desired result has been achieved. The achievement of the results requires more detailed planning of tasks or activities, which can again be done as a team activity once the major milestones have been agreed. It is important that this logical sequence is followed as there is little point in planning details if the principles haven't been agreed. The approach could also be considered as a 'pull' system, similar to Kanban in manufacturing, where information is generated to match the needs of an immediate deliverable rather than just generating engineering information, whether or not it is needed.

The term (vertical) value stream mapping is also used to describe this approach, although this terminology can lead to confusion with value stream mapping as a means of identifying waste in established and relatively stable processes.

PROJECT MILESTONES – Development of Technology Programme Management Method							
Result paths: A – project management B – user selection & gathering of needs C – trials of management method D – management method E – exploitation plan						**Project:** Development of Technology Programme Management Method	
Planned Date	**A**	**B**	**C**	**D**	**E**	**Code**	**Milestone**
17-Oct-16	O					1	Project start – agreeing the overall approach between the parties
08-Nov-16	O					2	When the team has been mobilized and the project launched
30-Nov-16		O				3	When the trial organisations have been identified and agreed to participate
30-Nov-16		O				4	When the process for understanding the needs of trial organisations has been agreed
15-Dec-16		O				5	When data has been gathered about the needs of trial organisations
31-Dec-16				O		6	When initial design of the management method has been finalised
15-Feb-17			O			7	When the system has been tested with the trial organisations
28-Feb-17				O		8	When the system has been redesigned based on the trials
15-Mar-17			O			9	When the final system has been re-tested with some of the trial organisations
30-Mar-17	O					10	When the work has been written up
30-Mar-17					O	11	When the plan for subsequent development has been completed
31-Mar-17	O					12	When the project has been completed

Figure 5.5 Milestone chart example.

A milestone-based approach is commonly used in early-stage software projects. The milestones are sometimes described as 'scrums' and the intervening activities as 'sprints', whilst the overall approach is described as 'agile project management'. Chapter 12 provides some more information on this topic, as does Ref. 5.3.

5.9   Risk Management

All projects carry risks which can undermine the achievement of project goals. The purpose of risk management is to bring these risks out into the open so they can be identified and then addressed. Risks which are hidden through ignorance or fear will come back later, causing problems and delays. In this context, the section of this chapter is concerned with *project* risks, rather than *technical* risks which are the subject, in detail, of Chapter 7. However, a project risk could be an inadequate technical risk management process.

Figure 5.6 illustrates how a project risk management process might be tracked – somewhat simplified. The steps are:

1) Identify a potential risk, with a tracking number assigned if the project is complex.
2) Assign the risk to an owner, who will have responsibility for managing it.

PROJECT MANAGEMENT RISK REGISTER – PROJECT XXX											
	INITIAL RISK					ACTION PLAN	OWNER	RESIDUAL RISK			
Risk Number	Description	Risk owner	*Impact*	*Likelihood*	*Risk Factor*	*Plan for managing the risk*	*Responsible for implementing the action*	*Remaining risk after action completed*	*Impact*	*Likelihood*	*Risk Factor*
1	Inability to recruit staff of the required experience	AB	5	3	15	Survey in-house possibilities, contact associates and known contractors					
2	Over-spend of the planned budget	MN	4	3	12	Compile detailed budget and discuss in detail with participants. Monitor costs weekly.					
3	Lack of clarity over the roles of the partners	CD	4	2	8	Understand the interests of each partner, document and share/discuss.					
4	Inability to find industrial partners for the work	CD	4	5	20	Document the aims of the project and circulate to account managers					
5	Lack of availability of specialised test resources	AB	5	3	15	Identify and book facilities as an immediate priority. Look for alternatives where there are problems.					
6	Timing over-run	MN	3	4	12	Complete timing plan and ensure everyone's support. Monitor progress weekly.					

Figure 5.6 Example of simple risk management matrix.

3) Assess the severity of the risk as originally identified – usually by estimating the impact and likelihood on a scale of 1–5 and then multiplying these two factors.
4) Compile an action plan and who will undertake the required actions.
5) Assess the current, or residual, risk once actions have been undertaken.

Risks severities using this method are usually categorised as 'high', 'medium', and 'low', as illustrated in Figure 5.7 and the aim obviously is to undertake actions, which move all identified risks into the 'low' category. Most risks may start as 'high' or 'red'.

Different projects have different risks, and some are inherently more risky than others, especially if they are of an exploratory or experimental nature. However, there are a number of common risks affecting many projects:

1) Availability of people, especially in the early stages of projects
2) Overoptimism concerning timing – especially assuming that nothing will go wrong
3) Overspending
4) Availability of physical resources – issues with offices, workshops, analytical or modelling facilities, especially in the early stages of projects
5) Lack of clarity about project objectives
6) Lack of clarity of roles of either individuals or organisations, especially when the project involves collaboration between different parties
7) Failure to understand customers' needs

Dealing with the above will never guarantee a problem-free passage for every project but it will reduce the likelihood of problems and will bring them out into the open more effectively and earlier. It should be remembered that post-mortems on failed projects have usually shown that problems did not come out of the blue but were known about and could have been acted upon if only they had been flagged up at an early enough stage.

Figure 5.7 Risk categorisation.

Likelihood					
5	5	10	15	20	25
4	4	8	12	16	20
3	3	6	9	12	15
2	2	4	6	8	10
1	1	2	3	4	5
	1	2	3	4	5

Severity

5.10 Resource Planning

Estimating and planning the manpower needed to undertake the work is an integral part of setting up a new technology or product development programme. It is required first to provide a cost estimate of the manpower element of a project and second to secure the people needed for the work itself. There are no first principles from which to make these estimates, which must therefore be derived from experience. The ideal experience would be similar, and fully completed, projects. Failing this, work in a similar field can be used, scaled up or down to suit the situation.

There is a natural tendency to be optimistic when making manpower or timing estimates; budgets are never quite enough and time is always short, so there is pressure to underestimate and assume that nothing major will go wrong. Hence, when using historic data, the actual resource consumed should be used – not what it should have been if everything had gone well. Figure 5.8 illustrates what a resource plot versus time might look like for a typical project. The plot of cumulative resource consumed versus time invariably follows an S-shaped pattern, and they are widely known as S-curves.

When estimating manpower needs, which are inextricably linked to the time allowed for the project, there are three particular points to bear in mind:

1) The divisibility of the tasks – some jobs need to be undertaken by one person, whereas others can be subdivided among several to achieve a faster result.
2) The effort to train, manage, and coordinate the team increases with its size.
3) Some tasks, such as durability tests, take a fixed length of time independent of the level of manpower applied to them (pregnancy generally takes nine months no matter how many people are assigned to monitor the progress).

These points present practical constraints on what can be achieved within a given timescale – see Ref. 5.4 for more discussion of these points. Early-stage tasks usually rely on the work being done by a small number of skilled individuals, and they cannot be shared out until a certain amount of progress has been made and a brief can be given to subsequent team members. If this early effort is held back by difficulties in manning the initial stages of a project, a frequent problem in most organisations, it is very difficult to catch time back by overmanning in subsequent phase of work.

Larger engineering organisations often forecast their future resource consumption looking ahead several years, resulting in a plot of the form of Figure 5.9 below showing the aggregation of the projects 'on the books'.

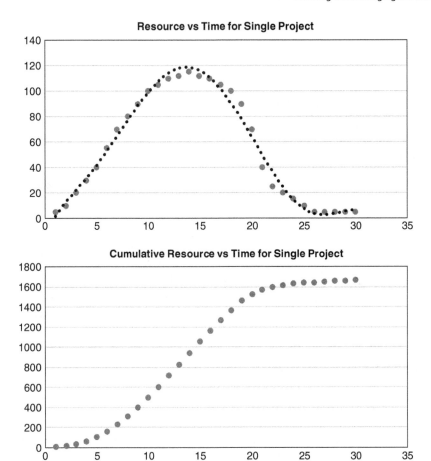

Figure 5.8 Resource plots for a typical single project.

Forecasts of this nature should include some level of support to existing products, often taking some 20% of effort. Graphs of this type usually show a deficit of resource in the short to medium term as new projects pile up, resulting in a 'bow-wave' profile to them. The consequence of this form of resource profile, as noted above, is that new projects either start late or are undermanned in their early stages, also resulting in delays.

One final point: once the work has gained momentum, it may be the case that adding new manpower does not necessarily speed up the work. The author has experience of one large and complex project where the work accelerated when manpower was reduced – too many people were getting in the way of each other.

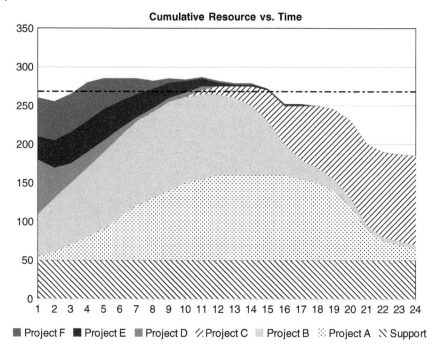

Figure 5.9 Typical forecast of future resource requirements.

5.11 Project Contingency

As well as estimating the most likely cost and timescale of a project, it is good practice to estimate a contingency amount representing, for example, the upper and lower boundaries of cost and time parameters. Management processes for large, complex projects invariably include contingencies along with procedures for releasing them and measures of contingency usage versus the level of completion of the project.

Clearly, early-stage projects are more difficult to predict than those conducted at a later stage where there is more detailed definition of the project content. Hence, contingencies expressed in percentage terms will be higher at early stages (but lower in absolute amounts because the projects are smaller in value), reducing as the work becomes more clearly defined. These contingencies cover both the effects of estimating errors (e.g. costs forgotten) and problems encountered.

Several standards suggest the level of contingency that should be applied, in money terms, to projects at different stages of development. In the Association for Advancement of Cost Engineers' standard, for example, a 'Class 5' estimate, the earliest, includes overrun boundaries of +30% to +100% and underrun boundaries of −20% and −50%. The corresponding boundaries for 'Class 1' estimates are +3% to +15% and −3% to −10%.

Contingencies are almost always expressed in monetary terms. On later-stage delivery projects, contingency sums can be spent to find ways of recovering the time effects of problems and thus minimise time overruns. In this situation, money can be converted into time. On early-stage projects, the number of people involved is much smaller, and there is more specialisation so there is much more limited scope for time recovery by overmanning beyond simply putting in more time. On early-stage projects, there is therefore an argument for regarding the percentage overrun boundaries as being relevant to both time and cost.

An active process of contingency identification and management is always a good way of dealing with the sensitive and inevitable subject of how to deal with project problems.

5.12   Organising for Projects

All projects, no matter how small, are organisations in their own right with very specific purposes. They may need to draw in a very wide range of skills, in addition to the technical and engineering skills often forming the core of technology or product development projects. For example, they will need manufacturing, marketing, financial, and commercial skills. Some will be required full-time and others for part of the time, and perhaps for only part of the project.

Given this background, there are then two ways in principle in which organisations can arrange their projects, with a wide variety of hybrids of the two models:

1) *Functional model.* The project's participants remain part of their 'home' discipline but are deployed onto the project for part, or all, of their time.
2) *Project model.* The participants are transferred, including physically, to a project for the bulk of its duration.

Both models have their advantages and disadvantages. In the case of a start-up organisation, the company may, in effect, be the project so no choice is available. With larger companies there is a choice. The relative merits of the two models are (Figure 5.10).

Where companies have large, multiyear and complex projects, the functional model is rarely effective and most companies in this situation adopt the dedicated project model. However, the disadvantages are real and the functional model can be made to work on relatively large projects if attention is paid to personal interaction and team development. This often goes against the grain with engineers who like to be given a task and be allowed to get on with it, undisturbed.

	ADVANTAGES	DISADVANTAGES
FUNCTIONAL MODEL	Minimises organisational disruption	Base department's priorities may not reflect project priorities
	Brings the organisation's full professional expertise into play, easier to transition people into and out of project	Leadership and coordination of projects is more difficult and there may be friction between projects and functions
	Employees don't leave their base and specialist career ladder is stronger	Participants do not identify as strongly with their projects
	Easier transition at end of project	Participants feel they have multiple leaders
	Learning transferred from one project to another	
	Greater opportunity for standardisation	
PROJECT MODEL	Provides project leader with a clearer brief and strengthens his position, making it a more attractive career path	The company's full expertise may not be available to the project
	Team communication and cross-functional working improved, especially if the project team is co-located	Technology priorities may be driven by project managers who are nonexperts
	Clearer priorities, project loyalty and focus for the participant	Project priorities make drift away from company priorities
	Faster overall delivery of results	One project may repeat the problems of parallel projects
		Less easy to standardise parts or platforms
		More difficult transition at end of project – may discourage people from joining a project if the organisation has a history of handling the end of projects poorly

Figure 5.10 Advantages and disadvantages of different organisational models for projects.

Many times, however, some form of hybrid is used, sometimes referred to as a 'matrix' organisation. Examples include:

1) Strengthening the project axis by having a permanent project organisation with heavyweight project managers but retaining the functional organisation and its advantages
2) Having a permanent project core team but deploying other resources into the team when needed
3) Having a permanent project team dealing with a range of related projects rather than one, in effect taking responsibility for a business stream rather than a project

As such, they have a degree of independence from the host organisation, creating both advantages and disadvantages.

5.13 Monitoring Small Projects or Subprojects

All projects, especially those with a strong element of learning or exploration, need to be monitored and kept on track. By their nature, such projects throw up issues rather quickly. This does not mean there must always be frequent and radical changes of course. Rather, the nature of development projects means that effective solutions to issues need to be found frequently and quickly as work progresses. This thinking applies to all projects, either to small projects in their own right or to the individual elements of larger projects.

The most effective way of doing this is the daily or weekly 'stand-up' meeting where all the participants get together to discuss progress, problems, and immediate action plans. Given the interactive nature of development work, such meetings are also opportunities to raise emerging or partially understood issues.

A project board, or a 'project on a page' document, can be used as the focal point for such meetings. Figure 5.11 is an example of such a board.

As well as background facts about the project or subproject, the board has three principal areas:

1) Graphical presentation of the timing plan and resource and cost consumption versus time, updated as the project proceeds
2) Summary of the current risks and their red/amber/green status
3) List of current tasks and issues, including responsibilities

The approach is informal and relies heavily on personal interaction but is very effective. The concept can be extended to creating a project war-room (*Obeya* in Japanese) with all relevant project information on display and placing the emphasis on visual management, including reporting to senior management, rather than relying on written reports. It is one of the approaches used by

PROJECT NAME			OVERALL PROJECT HEALTH		R/A/G	
PROJECT REFERENCE			PROJECT AIMS *(text)*	RISKS	RISK STATUS	
TEAM:				1.	R/A/G	
TIME PLAN *(graph)*	COST PROFILE *(graph)*		RESOURCE PROFILE *(graph)*	2.	R/A/G	
				3.	R/A/G	
				4.	R/A/G	
				5.		
ISSUES/TASKS						
NOT STARTED	*IN-PROGRESS*		*COMPLETE*			
1.						
2.						
3.						
4.						
5.						
Responsible for board:						

Figure 5.11 Project monitoring board.

Toyota and is an element of Toyota's rapid product development systems. Something similar is used in the field of software development where the project room is sometimes known as an 'information radiator'.

These approaches can also be used for elements of large projects, but these will almost certainly need a more formal, documented approach for senior management review.

5.14 Approval and Formal Monitoring of Large Projects

A common way of overseeing large projects is through a 'stage-gate' or 'phase-gate' process. In this approach, a project is divided into a series of phases with formal evaluation gates at the end of each phase. For the project to proceed to the next phase, it must demonstrate that the pre-agreed criteria for the previous phase have been met; it might also be necessary to show that the resources are available for the next phase (money, people, and facilities).

This approach is used in both product and facility-related technologies and has its origins in the mid-1980s through the work of Professor Robert G. Cooper, who studied the practices of successful companies at that time, many of whom were using already some form of gated processes – see Ref. 5.5. It is used widely in the oil, gas, and process industries, as well as in product-related organisations.

In the context of developing a new technology, a stage-gate process of this type only becomes relevant when the technology has entered the 'product development' phase of its maturity, i.e. there is a committed plan to commercialise the technology through a launch product. The technology readiness would therefore be at approximately TRL5 (technology readiness level) when being scoped into a project and TRL 6 when the business case is agreed and a commercialisation project, including a new technology, is launched.

A five-phase model is quite common with, for example, the following phases:

1) Defining the scope of a project
2) Developing the business case and agreeing the go-ahead
3) Development of the product itself
4) Validating the product
5) Launching the product

The criteria for evaluating whether a gate has been passed can include:

1) Technical development status
2) Manufacturing development status
3) Marketing readiness
4) Commercial factors
5) Business case

A tabular summary of a potential stage gate structure is tabulated in Figure 5.12.

There is also the question of the rigour, which is applied to any evaluation – tasks that are allegedly 90% finished are a particular problem. A distinction can be drawn between essential criteria, without which progress cannot be agreed, and desirable criteria, where some level of tolerance might be applied. In fact, some weeding out of nonessentials is frequently required.

The process can be applied in a very structured way, with preset deliverables, predefined criteria, and specific project decisions categories such as: proceed, resubmit, hold, or cancel. Hence, the process does require some independence of thought, a willingness not to fudge issues, and a preparedness to cancel projects. The approach is best suited to large companies with multiple projects, but the same disciplines can be used in small and developing companies even though reviews and decisions may be made by a limited number of people without the support of teams preparing large quantities of data.

Stage Phase Gate		Technical Status	Manufacturing Status	Market Readiness	Business Case
1	**Scope**	Technology developed to TRL5 maturity level.	MRL 3: Manufacturing proof-of-concept developed. Experimental hardware or processes created. Materials or processes characterised for manufacturability and availability. Initial cost projections made. Supply chain requirements determined.	Product and market proposed, competition evaluated	Outline business case developed
2	**Business case,** including full, timed & committed project plan	Technology developed to TRL 6 maturity levels. Product defined and evaluated against market & customer requirements	MRL 4: Series production requirements identified. Processes to ensure manufacturability and quality in place and sufficient to produce demonstrators. Manufacturing risks identified for prototype build. Cost drivers confirmed. Design concepts optimised for production	Market analysed, product tested through customer discussion, competition evaluated, launch product agreed, channels to market identified	Full business case including investment, sales, margins, cash flow and payback/return data, risk analysis
3	Full product definition	TRL 7 – launch product defined in detail and prototypes built	MRL 5/6: Prototype manufacturing processes is place, production process defined, supply sources and lead times known, detailed cost analyses completed	Market and sales plans developed in detail	Business case updated using latest product and market data
4	Product validation	TRL 8 – launch product validation programme completed	MRL 8: production system in place, facilities proven, pilot runs undertaken to prove capability	Sales and marketing channels and commercial arrangements in place. Product trialled with customers.	Business case updated using latest product and market data
5	Launch	TRL 9 – product cleared for launch	MRL 9 – manufacturing system cleared for volume production	Product launch process in place with sales network, promotional programme, trained staff, and financial support	Monitor customer acceptance & business performance through launch process

Figure 5.12 Outline of stage-gate model.

5.15   Project Management versus Technology Maturity Assessment

This chapter has dealt with the management of technology and product development projects. It may appear to overlap with Chapter 3, which is concerned with evaluating the maturity of technology. There is, however, a clear distinction between the two. Project management relates to the tasks, costs, and timescales for achieving prescribed objectives, which could be technical (such as achieving defined technology maturity level) or could be business-related (such as launching a new product). Technology maturity development, on the other hand, is concerned with achieving a certain level of technology readiness, as measured by TRL scales and implying a certain level of technical risk reduction, with project management as a mechanism for overseeing this process. A good understanding of technology maturity is a prerequisite for competent project management.

5.16   Concluding Points

New technologies and new products are generally delivered through projects, that is, time-bounded activities with specific objectives. As technology advances in maturity, the projects for delivering new technology or new products become more structured, more complex, and more commercially focused, with an expectation by investors that results will be achieved. Whilst agile or highly responsive attitudes can dictate the speed with which projects are completed, the basic disciplines of project management are valid at all phases of development – the 'fuzzy front end' through to multimillion-pound commercial projects – and should be understood by all engineers whether they are working at research-oriented or delivery-oriented activities.

References

The goal-directed approach was first developed in Norway and is not very well known but is well thought through. Its principles are quite similar to "agile" methods (see below):

5.1 Andersen, G. and Haug, K.P. (2009). *Goal Directed Project Management*, 4e.

This book is one of the classics of product design and contains a lot of very practical information, although the details have been overtaken by time and the work precedes the widespread adoption of CAD and other digital methods:

5.2 Pugh, S. (1991). *Total Design*. Englewood Cliffs, NJ: Prentice Hall.

For a straightforward overview of agile methods, this book is a good introduction:

5.3 Layton, M.C. (2012). *Agile Project Management for Dummies*. Hoboken, NJ: Wiley.

There is some interesting material on how to plan and estimate the resources needed for software, and technology, projects:

5.4 Brooks, F.P. Jr. (1982). *The Mythical Man-Month – Essays on Software Engineering*. Reading, MA: Addison-Wesley.

Robert Cooper was one of the first to publish material in the 1980s about stage-gate methods, and his work is always well researched:

5.5 What's Next?: After Stage-Gate Research-Technology Management January – February 2014, pages 20–31, Reference Paper #52 by Robert G. Cooper

6

Developing the Concept

6.1   Introduction

Reference was made in Chapter 2 to the tension between two fundamental aspects of the engineering process:

- The fact that it is essentially a creative activity, developing new ideas to solve problems and to improve people's well-being. As such, it is also the source of the innovation that drives business and economic growth in an industrial context.
- On the other hand, there is an element of conservatism in that those new engineering solutions must be reliable, robust, and not create harm, danger, or adverse environmental impact. In this respect, the world is becoming increasingly critical and risks or problems must be identified and fully overcome before a new technology or product is launched.

This chapter is concerned with the first of these points – the development of new concepts. This stage of work brings together three parameters: future market needs, possibilities in terms of new technology and engineering, and economic viability. A constant balance has to be struck between these competing requirements, as illustrated in Figure 6.1.

As an essentially creative process, it is not one which can be readily codified or proceduralised. However, the concept of good practice does apply and business school research (Ref. 6.1) into this front-end process has given some useful guidance, which is worked into the material below. The emphasis here is on early-stage work, ahead of the deployment of relatively large-scale engineering teams undertaking the more detailed stages of commercial delivery, which is covered in Chapter 9.

Managing Technology and Product Development Programmes: A Framework for Success,
First Edition. Peter Flinn.
© 2019 John Wiley & Sons Ltd. Published 2019 by John Wiley & Sons Ltd.

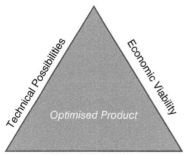

Figure 6.1 Balancing market, technical, and economic needs.

6.2   Key Elements of the Process

The first element of the process is the generation of ideas that might have some useful purpose in the eyes of potential customers. The ideas could come from any of the following:

- Company engineers or salespeople, who understand the products currently being sold and can see their shortcomings or their scope for improvement
- Market feedback, warranty, and complaint information, and customer surveys
- Long-range technology forecasts – such as the technology roadmaps described below
- New recruits to companies, bringing ideas from other sectors of industry
- Research engineers who are developing new technologies in university or corporate laboratories
- Start-up or spin-out companies, often linked to a university
- Suppliers to the company, noting that supplier companies are given far more responsibility and involvement than was the case in former times
- Less likely but not unknown, private individuals in their garages and workshops

Spillover from one sector to another is one of the most fruitful sources of innovation. It often occurs through exchange of ideas, interaction, and sharing. The ideas mix and mutate, resulting in new versions of established ideas (Henry Ford admitted he had invented nothing new. He said he 'simply assembled the discoveries of other men behind whom were centuries of work'). The pace of such change has definitely accelerated as a result of greater personal mobility (both in terms of travel and movement from one employer to another), better communications and a conscious effort on the part of some governments to promote knowledge transfer networks.

The second element of the process is knowledge of the marketplace and the spotting of gaps or opportunities that could be of interest to customers and that could form the basis of future business opportunities. There may also be the possibility of creating new markets although this is more unusual.

The third element is then the iterative synthesis of these ideas, in the context of the market opportunities and their economic feasibility, to form a workable concept.

At this point, it is worth drawing distinctions between different levels or forms of innovation: they could vary from the modest, incremental improvements (which is the most frequent) through to radical, break-through or disruptive changes. The former are relevant to existing market structures and represent the progressive, and probably annual, improvements that consumers expect of their cars, washing machines, or mobile phones. The latter are the less frequent but radical changes that create completely new markets such as those that did not exist 30 years ago for personal computing or mobile phones.

In reality, new ideas will sit on a spectrum with these as the two extremes, as shown in Figure 6.2.

A further dimension is what is sometimes referred to as 'adjacent' development, where an existing technology or solution is applied to a different market sector. It is easy to underestimate the effort required to move into a different market where the operating environment and usage patterns can be quite different. For example, automotive solutions have often proved quite unsuitable for rail applications as a result of the much harsher mechanical and electrical environment (such as 25 kV overhead lines), the heavier and rougher use in the rail sector, and the 30- to 40-year life requirement. Nonetheless, spillover between sectors is the basis of many innovations, provided that problems of the type indicated can be overcome.

Research has suggested (see Ref. 6.2) that the optimal split between these forms of improvement in a large and successful company, with a balanced portfolio of technology developments, is:

- Incremental – 70%
- Adjacent – 20%
- Breakthrough – 10%

The corresponding financial returns on successful projects are then the opposite of these – 10%/20%/70%, but of course the 70% return on breakthrough innovations carries a much higher risk. In fact, breakthroughs, or 'disruptive' changes cannot really be achieved to order and may not be recognisable as such

Figure 6.2 Spectrum of new ideas.

changes for some time after their invention and often outside the field for which they were originally intended, lasers being a good example of this.

6.3   Technology Roadmapping

One of the policy tools used by large corporations, public bodies, and research funding organisations is technology roadmapping. It is used to guide their early-stage investments. The form described here was developed originally in the 1970s by Motorola as a means of overseeing and managing its considerable portfolio of technologies in the field of electronics. Since then, it has been widely adopted by the range of organisations noted above. It is also relevant to small and medium-sized companies, either for managing their own work or as a source of guidance about the trends in markets and technology.

The approach is used to navigate the complex field of new technological opportunities by mapping out the opportunities and help choose which options to follow. It also provides a framework for reviewing decisions as new circumstances arise or as new information becomes available.

The results of this planning work are usually presented in the form of a chart with time as the horizontal axis, covering anything from a 3-year to a 20-year horizon. Figure 6.3 is a typical example. Planning is best carried out in a facilitated workshop environment and there are some highly structured methods for running workshops of this type. This approach enables the combined expertise of the organisation, its partners, and independent experts, to be brought into play.

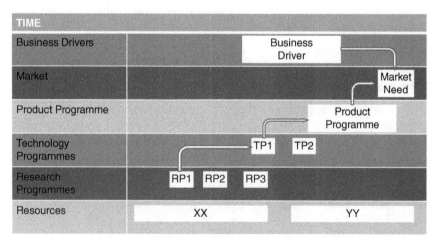

Figure 6.3 Technology roadmap example.

The chart typically would have between three and six interconnected levels covering, from the top downwards, such topics as:

- Business drivers
- Market requirements
- Products and services
- Technologies
- R&D programmes
- Resources

The point of the process is to link the business and market drivers, at the higher level of the chart, to the technology programmes placed lower in the chart. In summary, the objective is one of aligning technology investments with market needs. As with all processes of this type, much of the value comes from the process of discussion, rather than the end result, although the highly visual form of roadmaps also makes them excellent communication tools.

This is especially relevant to a single-company environment and can drive key decisions such as when a new technology will be sufficiently mature for it to be launched on a new product development programme. The weakness of the process may come when a new technology creates a new market.

Roadmaps are also a favourite of public bodies, although they are often presented in narrative, rather than pictorial, form. A search of the internet will produce documents on every new technology where there is government interest: solar power, nano-technology and nuclear power being just three examples. Roadmaps also often define the areas where research-funding bodies will, or won't, provide funding.

For all organisations, small or large, technology roadmaps are a good, visual, and easy-to-understand way of communicating the direction of the organisation in the field of technology and products. More can be found in Ref. 6.3.

6.4 Open Innovation

The points above consider innovation as a largely internal activity, carried out within the boundaries of a company and its partners. Looking more broadly, the term 'open innovation' was coined by Henry Chesbrough (see Ref. 6.4) in his 2003 book describing new models of innovation, which could potentially accelerate the pace of development of new products and services and involve a wider community in the process. It is sometimes presented as a paradigm shift, i.e. a fundamental and revolutionary change of practice resulting in radically different results. In reality, it is more of a progressive trend over a long period in which organisations place more reliance on external sources of ideas, rather than depending entirely on their own resources.

The trend is manifested in a number of ways. Engineering companies, for example, are placing more reliance on their suppliers and on university collaborations, where some large corporations designate certain university laboratories as the sources of defined low-technology readiness level (TRL) technologies. Pharmaceutical companies are using start-up or spin-out companies as the initial source of new products.

The software industry has progressed furthest in this field, some software being developed in an entirely collaborative, or open, manner within a defined framework.

However, there are competing trends on this topic: more open models of innovation are being used, but at the same time, intellectual property (IP) protection is being tightened up by many organisations.

6.5   Concept Development

Moving onto the work itself, the focus of these early-stage activities is the development of a defined concept, which can then be evaluated for its merits and its feasibility as a potential business proposition. It is too early at this stage to speak of a full business plan, but the elements of ideas, technology, manufacturing feasibility, and value to the market must be there.

The process can be characterised as a series of divergent/convergent iterative loops, as shown in Figure 6.4.

In the first divergent area of development, markets are understood and ideas are generated, analysed from an engineering and manufacturing perspective, and discussed with customers. After reworking them, they are pulled together as complete, but draft, proposals to see whether they make sense as a whole. There could be some level of formal review at this stage with people outside the development team, or the team itself might reach its own conclusions and start another round of development.

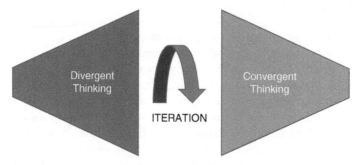

Figure 6.4 Divergent/convergent iterative loops.

VALUE PROPOSITION MAP

	SUPPLIER	CUSTOMER	
BENEFITS	Beneficial features	Positive needs	NEEDS
	Avoiding features	Problems to be avoided	

Figure 6.5 Value proposition map.

Documenting the work at the end of each loop has some important advantages. First, the discipline of writing things down enforces organised thought and the tying up of loose ends that might otherwise be left unresolved. Second, the proposal will have to be sold to someone: a senior person in an organisation, a product policy or technology development committee, or an external investor, who will require more than verbal information. Third, it keeps track of what has been done so that the learning can be used in later stages of the project in question or on other, future projects.

At a relatively early stage, it may be helpful to draft a value proposition in the form outlined in Chapter 4, including a completed example, and repeated as shown in Figure 6.5.

As noted previously, the most important element of value proposition development is the headline statement, which acts a guide and focal point for the new technology or product. This statement can also be built on by adding, in more detailed product definition material, further information about how customer needs will be met, as indicated in Section 6.11, 'Linking Detailed Design to Customer Needs'.

As implied, these activities are best conducted in a small-team environment using experienced leadership well connected to the marketplace and to the senior management of the company, in the case of a corporate environment. More thoughts about development teams, their composition, and their leadership are given in Chapter 9.

The process described above could take an incremental improvement to a technology readiness level of $\sim$TRL 5 and a more radical technology to $\sim$TRL 3. The more radical the proposal, the more iterations will be required and the more analysis, physical testing, and market discussions will be needed. The aim should be to achieve a position where the proposition can be evaluated sensibly from a broader, business perspective, and then a decision taken whether to proceed further with more detailed, and more expensive, stages of work. For this to be successful, there must be an open flow of ideas and the ability to test these ideas with the market.

6.6   Industrial Design

Although not stated specifically, the concept development process outlined above is usually understood as describing the early stages of 'engineering design', with the emphasis being on technical performance and function. It is also the period when the principles of 'industrial design' should be applied. This latter term is generally used to describe development of the aesthetics, styling, appearance, ergonomics, and feel of the product, as opposed to its engineering development, although the boundaries between the two is not hard and fast.

Different products will strike different balances between engineering and industrial design. Factory equipment, for example, will emphasise engineering function whereas domestic appliances will have much more emphasis on creative industrial design as defined here. Cars have long struck a balance between styling and function. Devices such as mobile phones, tablets, and laptops have become almost an extension of an individual's persona and must therefore have the functionality, feel, and appearance required for this role.

Development of the industrial design of a product can involve an interesting dialogue between the engineer and the industrial designer, often people with rather different personalities and outlooks on life. But it is not a 'zero sum' game. Whereas there is always some trade-off between industrial and engineering design, the best products achieve the best of both worlds, performing well and being aesthetically appealing – 'form follows function'. Industrial design is always very user-centric, an important element of product success, as noted below.

The critical point is that industrial design is an important element of concept development, and the principles to be adopted on a new product should be built in from the outset and not as some afterthought.

6.7   Key Success Factors

The drivers of new product success have been researched by a number of academic organisations (Ref. 6.5), and the three most important factors in this respect have been shown to be:

1) The superiority of the product in terms of the features it embodies and how the product then compares with the competition
2) The extent to which customer needs have been investigated in detail and then built into the product
3) The amount of effort invested in early-stage product development and the thoroughness with which this early work has been undertaken

None of these points comes as a surprise but they do reinforce the importance of investing generously in those early stages – a point made in Chapter 3 when discussing NASA's findings and their findings are worth repeating. Their historical cost analyses had shown that there was a direct link between the success of a technology and the amount spent on its early-stage development – the more the expenditure the less likelihood there was of subsequent cost and timescale overruns.

6.8 Identifying and Meeting Customer Needs

As already indicated, customer input is critical to the development of new concepts. The 'voice of the customer' is a common term in many areas of business, defined by American Production and Inventory Control Society (APICS) as the 'actual expressed desire for product functions and features'. And there are many ways of establishing what customers would like.

A widely used approach in this respect is the model developed by Professor Noriaki Kano of the University of Tokyo and winner of the Deming Prize for individuals. In his work in the 1970s and 1980s, he drew attention to different aspects of customer satisfaction, as summarised in the model in Figure 6.6.

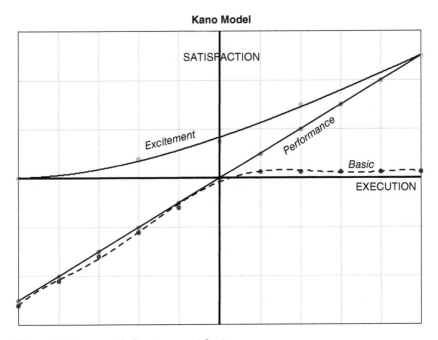

Figure 6.6 Kano model of customer satisfaction.

The model plots customer satisfaction as a function of the standard of product execution and draws a distinction between three groups of attributes:

1) *Basic.* These are the standard features that customers expect of a product and take for granted, whether they explicitly specify these features or not. For example, we all expect our cars to start at the first attempt, even on a cold morning. This is not something we would mention to a car salesperson when describing desired features, but it would be highly dissatisfying if the new car did not have this capability. 'Basic' requirements do not cause satisfaction or additional satisfaction, as such, but will cause dissatisfaction if not met. Identifying basic features is not always easy as customers tend to assume that they will be present.

2) *Performance.* These attributes provide more satisfaction the better they are executed (or less satisfaction if executed poorly). For example, a light travel bag causes satisfaction and a heavy bag causes dissatisfaction, other things being equal and assuming both function properly. Performance attributes are keenly watched by customers and form the basis of buying decisions.

3) *Excitement.* These factors are unexpected and come as surprises to customers who didn't expect them. They might be categorised as 'unique selling propositions'. Backup or reversing cameras on cars were in this category some time ago but are now regarded as standard – excitement factors quickly lose their edge as everyone adopts them.

The usefulness of this model, beyond interest or curiosity, lies in how it might guide data gathering from potential customers.

6.9 Customer Data Gathering

Understanding customer needs and drivers is critical to the early stages of product development and studies have shown that one of the most frequent causes of new product failure is poor understanding of customer needs. Early-stage technology and product development is usually perceived as the domain of engineers, who, to be fair, are often the sources of new ideas. However, if the marketing input is weak, it is easy to be carried away with new technology ideas that may have potential but may be somewhat off-centre from the customer's perspective. Once an idea is established, it can be difficult to shift, as natural defensiveness on the part of the originators comes into play. Marketing work is therefore not something that is used to check an idea after it has been expanded but should be regarded as an integral part of the early processes.

In terms of how information can be gathered, there are multiple ways in which this can be done. These are just some possibilities:

- In-depth interviews
- Extended observation of customers
- Clinics
- Social media
- Work directly as a customer, e.g. drive a delivery van
- 'Ethnography', involving in immersion in the consumer group under study to understand the culture of that group
- Testing of ideas through physical models, mock-ups, CAD models, and virtual reality
- Test runs with customers with prototype systems
- Data measurement to understand quantitatively the operating environment
- Warranty and complaints information from existing products

A point to make is that customers will find it difficult to react to radically new ideas that go beyond their current experience, and they will therefore be more comfortable responding to incremental improvements to existing products. It should also be noted that different customers have different needs and will use the same product in different ways. It is therefore important that a range of different customers are studied to understand the full spectrum of needs. This is particularly the case with engineering products where the range of applications and duty cycles can vary widely from customer to customer and from country to country. There may also be conflict of requirements where one attribute operates at the expense of another. In this case, developers may need to back one priority, and hence one customer, at the expense of another, recognising that no product can meet every requirement.

The actual process of customer observation and data gathering will always benefit from a structured and disciplined approach. Wherever possible, it should be tackled in the same way as a scientific or engineering experiment with:

- Defined methodologies and questions
- Trained staff or observers
- Structured data gathering and documentation of results
- Direct observation of use in the real environment (not through an intermediary 'expert')
- Range of situations observed
- Significant sample sizes
- Statistical analysis of results

The emphasis of work of this type is to understand what functions they want the product to perform rather than the features (i.e. the solutions) that may provide those functions and, critically, whether they are prepared to pay for those functions. It is best carried out by members of the development team itself, so they have direct experience, rather than a specialist group.

6.10   Who Is the Customer?

The foregoing material skates round the question of 'who is the customer?' In consumer markets, customers are numerous and varied but are typically single individuals. Hence, most products are targeted towards a specific customer group or segment. A balance has to be struck between numbers and fully satisfying particular customers' needs. The earlier references to niche-driven, differentiated, and cost-driven markets illustrate the different approaches that can be taken. A conscious choice has to be made as to what customer group to target and analyse.

In the case of business markets, clients tend to be fewer in number but more difficult to pin down. For example, a commercial vehicle may be owned by a leasing company or operated by a logistics firm but in the livery of a retailer. Who is the customer in this situation? In reality, multiple customers exist in this situation, all of whom have to be satisfied to some degree. In other business situations, the true customer may be less obvious. For example, the formal link between a supplier and a customer may be a purchasing officer. However, the choice of a supplier may be made by a design engineer who specifies a component on a drawing and who may not realise he is also choosing the supplier. Some thought is therefore needed in analysing these situations in terms of who makes the purchasing decision and who can say no.

6.11   Linking Detailed Design to Customer Needs

Having established in detail the customer's requirements, these then need to be translated into product engineering terms. (This is not a linear, or sequential, process with requirements cleanly preceding engineering work. In practice, the two will proceed in parallel).

An established way of doing this is through the 'quality function deployment' (QFD) structure. This approach was developed in Japan by Yoji Akao (also a Deming Prize winner) in the 1960s and 1970s and first used on shipbuilding projects in Mitsubishi's Kobe shipyard.

Called *Hoshin Kanri* in Japanese, the English translation of 'quality function deployment' is not the best. The concept refers to taking customer requirements (the ultimate measure of quality), translating them into engineering features (functions) and cascading them into the organisation (deployment).

At its simplest level, this linkage can be created in a tabular structure with customer requirements on the vertical axis and engineering characteristics on the horizontal. The two are then linked together – some customer requirements will relate to more than one engineering characteristic and vice versa. This linkage process is best accomplished in a team environment using a 'Post-it' note

			ENGINEERING CHARACTERISTICS						
			A	B	C	D	E	F	G
			QUANTIFICATION	QUANTIFICATION	QUANTIFICATION	QUANTIFICATION	QUANTIFICATION	QUANTIFICATION	QUANTIFICATION
CUSTOMER REQUIREMENTS	1	QUANTIFICATION							
	2	QUANTIFICATION							
	3	QUANTIFICATION							
	4	QUANTIFICATION							
	5	QUANTIFICATION							
	6	QUANTIFICATION							
	7	QUANTIFICATION							

Figure 6.7 Simple requirements/characteristics matrix.

style of workshop. Wherever possible, requirements and characteristics should be quantified, which may prove difficult initially but which can generally be built up over time.

This simple matrix is the core of the full QFD approach but could not be described as a 'proper' QFD analysis (see Figure 6.7).

There is a practical limit to the complexity of matrix (number of rows × number of columns) which can be handled by one individual or group. And, of course, it would be quite impracticable to describe a complex product such as a car, train, or aircraft on one single matrix. In this situation, a hierarchy of matrices can be constructed and the approach generally seems better suited to components or small systems rather than more complex products.

The more widely known 'house of quality' adds more structure, detail, and insight into the process of linking customer requirements to engineering characteristics, but at the expense of making the process look more daunting.

Illustrated in Figure 6.8, the process for developing this form of analysis is:

1) Document the *customer requirements* as above, including any regulatory requirements.
2) Assess the *priority*, which the customer places on these requirements and how the customer views the company's solutions relative to competitors, as a *competitive evaluation*, both on a 1–5 scale.
3) Enter the engineering characteristics (*product design requirements*) data both in descriptive form and as numerical targets (*target value*).

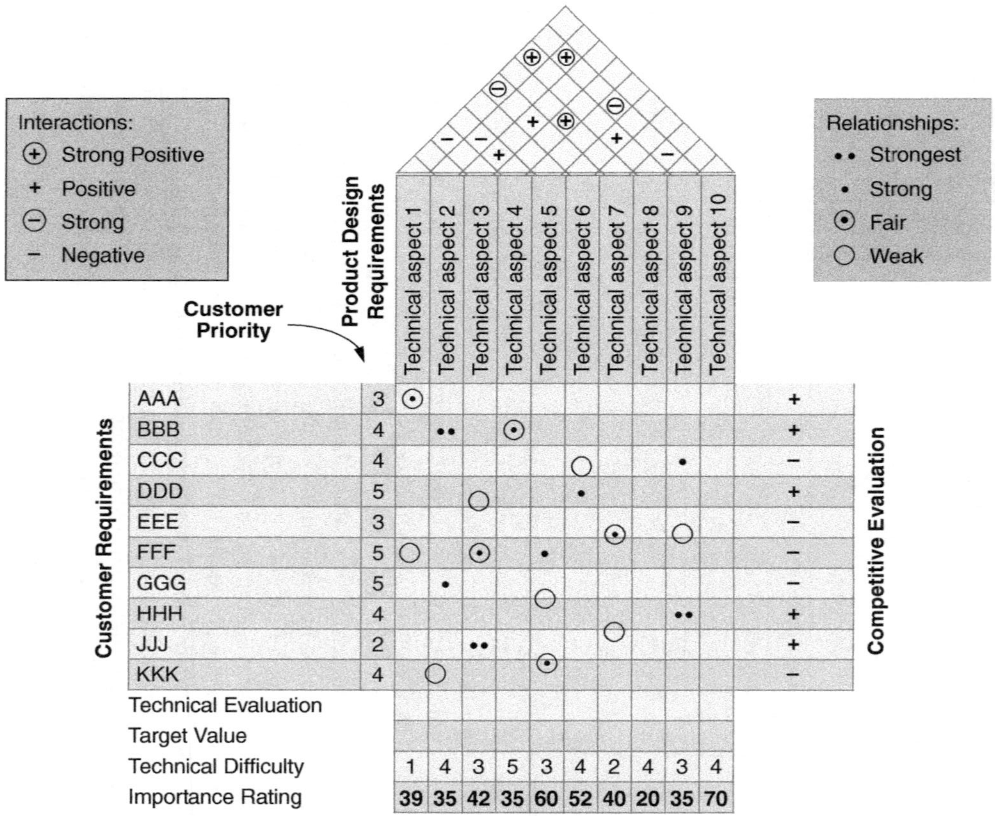

Relationships:
●● Strongest
● Strong
⊙ Fair
○ Weak

Interactions:
⊕ Strong Positive
+ Positive
① Strong
– Negative

Competitive Evaluation

Customer Priority

Product Design Requirements

Customer Requirements

Technical aspect 1
Technical aspect 2
Technical aspect 3
Technical aspect 4
Technical aspect 5
Technical aspect 6
Technical aspect 7
Technical aspect 8
Technical aspect 9
Technical aspect 10

AAA 3
BBB 4
CCC 4
DDD 5
EEE 3
FFF 5
GGG 5
HHH 4
JJJ 2
KKK 4

Technical Evaluation
Target Value
Technical Difficulty 1 4 3 5 3 4 2 4 3 4
Importance Rating 39 35 42 35 60 52 40 20 35 70

+ + – + + – – – + + –

Figure 6.8 Full QFD 'house of quality'.

4) Perform a *technical evaluation* of the proposed product versus the competition, again on a 1–5 scale. To do this well will need access to the products of the competition and, ideally, some form of teardown analysis.
5) Analyse the *relationships* between the customer requirements and product design requirements, distinguishing strong, moderate, and weak relationships. This helps to understand which engineering parameters might be adapted to provide a better match with customer requirements.
6) Analyse the *interactions* between the engineering requirements, which can be both positive and negative (e.g., there is usually a negative relationship between low weight and high strength).

As with many team-based activities of this type, the value of a completed QFD matrix lies more in the process for arriving at it, rather than the finished article. However, the completed matrix does represent one possible solution to a customer's needs which can then be adapted. All solutions have some level of compromise and trade-off, which are brought out in the matrix. Alternative solutions can be developed and analysed in the same way. Several competing QFD models can then be used to assess which provides a better balance of compromise and hence is more competitive.

The full QFD process is used quite widely in the automotive industry promoted, in particular, by the Ford Motor Company and may be mandated as part of a supply agreement.

6.12 Ensuring a Robust Design – Taguchi Methods

At this early stage of development, it is wise to invest in analysis that will make the product less sensitive to the method of manufacture or to environmental factors when in operation – how it can be put together in a way where it works readily and does not require constant 'fiddling' to make it function, or keep it functioning, correctly. The leading thinker in this field was Dr. Genichi Taguchi (1924–2012) who worked in both Japan and the United States. The author was taught his approach by his son, Shin Taguchi, who is still active in this field at the time of writing (see Ref. 6.6).

Taguchi's initial work came from involvement in the design and manufacture of telephone switching systems, which at the time of his involvement were complex electromechanical devices. His approach revolves around three concepts:

1) A loss function, which describes how losses to the manufacturer, customer, or society more generally increase as a parameter, such as a dimension or a voltage, deviate from their ideal value. His definition of quality is, '*the loss imparted by the product to the society from the time the product is shipped*'.
2) Development of the robustness of a design in the early stages of development as a form of off-line quality management. This is complementary to on-line methods such as inspection and statistical quality control.

3) The design of statistical experiments to investigate, efficiently and numerically, the two points above and thus provide hard data that can be used to reduces losses and increase robustness.

The loss function, illustrated on Figure 6.9, is at the heart of this thinking.

Taguchi's approach draws a distinction between traditional tolerance bands, where something is 100% acceptable if within the band and 100% unacceptable if outside it, and a more progressive, continuous approach where losses, or problems, increase the further the parameter is away from the ideal. Ford Motor Company had practical experience of this in the 1980s when they compared automatic transmissions made by themselves and Mazda to identical specifications. The components in the Mazda transmissions used up less of the available tolerance bands and had better whole-life costs despite being slightly more expensive to make.

Investigating the sensitivity of, say, the performance of an electro-mechanical mechanism or the output of a TV power supply, to parameters and variabilities within them is a way of understanding how the design and the processes can be set up to provide a reliable overall result. Multi-parameter studies can be used to investigate different sources of 'noise', or variation, ahead of production. These sources of noise could include, for example, the operating environment, human errors, or piece-to-piece variation. The term 'signal-to-noise' ratio is used to describe robustness.

He did a lot of work on 'design of experiments' (DOE) to find ways of conducting experimental analysis of situations where there are multiple variables but where there is simply not enough time or money to experiment by varying one parameter at a time, as would classically be done. Examples are available

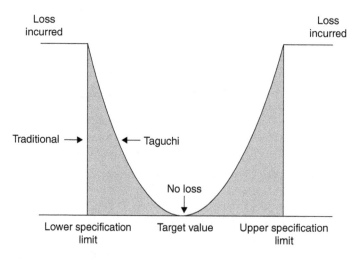

Figure 6.9 Taguchi loss function.

for such applications as ceramic tile manufacture, control valve performance and copier mechanism jamming.

Taguchi put in some considerable mathematical work to support his approach and there is some debate about his methods amongst the statisticians' community. It is probably fair to say that his approach has been used more as a conceptual tool ahead of production than as a detailed methodology. However, there are companies who use DOE as part of their development processes. The mathematics of experimental work can be complex but where a new concept design is proving difficult to put together in a way that gives reliable results, the experiments can be a powerful way of understanding the causes of variability and hence making the product more robust. A conceptual understanding of Taguchi's approach is a useful element of the engineer's armoury.

6.13 Technology and Manufacturing Development at the Concept Stage

The technology activities and outputs that take place during development of the concept include typically:

- Schematic or general arrangement CAD models or drawings
- Performance specifications
- QFD models
- Taguchi models
- Narrative descriptions and illustrations
- Mathematical simulation models
- Test pieces and laboratory test results on critical areas of the product
- Mock-ups
- Intellectual property searches and initial IP protection of critical areas
- Initial risk analyses and/or design failure modes and effects (FMEA) studies
- Initial cost studies

Manufacturing work will follow a similar pattern and could include:

- Initial manufacturing feasibility studies and outline process definitions
- Outline facility plans
- Test work on critical processes or process FMEA studies
- Some assessment of supply chain requirements

This early-stage technical work forms the foundation of future development. On the one hand, it develops a concept that will appeal to customers. On the other hand, it must identify critical issues and risks and do sufficient work on them to provide confidence that they can be overcome in subsequent project phases.

6.14   Economic Evaluation

Whilst the bulk of the work in developing a concept is market-led engineering, it is also important to be able to assess whether the proposal being developed will work from a financial point of view. At this stage in a development programme, rigorous financial analysis will be difficult but it should be possible to make some level of financial judgement. In principle, this requires estimates of sales volumes and prices, product cost, and one-off investment and launch costs.

In an established market or where an incremental improvement is being made, this can often be done by reference to existing costs or from experience of previous, similar projects. This will indicate whether a project makes sense or, if not, what should be done to achieve financial viability.

Analysis will be more difficult in the case of a breakthrough technology where there is no clear point of reference in terms of markets or prices. Costs can be built up with the help of suppliers. Volumes are more difficult, but any estimate should be very realistic – don't say '72 million cars are produced annually worldwide and if we can get on just 0.5% of those, that's 360 thousand units'.

Prices may come from analysing the market and judging what, logically, customers might pay to be attracted to the new product. A more rigorous approach is to work analytically through the product, function by function, to confirm that the customer will pay for it and to estimate its value. In the case of industrial products, this would be linked to the cost and commercial benefits of the new solution.

The method of analysis would then simply be to judge the profit from the sales against the original investment to establish whether there is a sensible payback period. Given that later stages of product development, after completion of the concept work, are increasingly expensive, it is important that this initial, simple analysis is based on realism. It is better to have a simple but realistic set of estimates than a more sophisticated analysis based on unsound assumptions. In later stages of a project, a timed, discounted cash flow analysis would be more appropriate. A word of warning at this stage: research (Refs. 6.7, 6.8) suggests that only 28% of new products meet their profitability targets and usually fail as a result of a poor match between what is being offered and what the customer will pay for.

6.15   Protecting Intellectual Property

Intellectual property (IP) is a collective term that is used to describe patents, design rights, copyrights, trademarks, and confidential information. New intellectual property is frequently generated during the concept phase of work; in fact, the generation of new ideas is the principal aim of the work and now is the time to protect it. Any drawings, diagrams, prototypes, and software code do create intellectual property rights as they are created and may be patented for protection. The scope for protection may be greater than realised in the sense that even new combinations of existing technology may qualify for protection – the ideas don't have to be radically new. However, if protection is to be sought, care must be taken not to disclose the ideas publicly before the protection process has started. Marketing pressures may encourage early disclosure at conferences or in discussions with potential clients to show that an organisation is innovative, but this could prejudice formal protection.

The various forms of IP protection include (Ref. 6.9):

- *Patents.* These protect novel inventions that can be used in a practical way. They are important to consider in relation to engineering projects as they provide a monopoly right that can be used to recoup R&D investment over a period of exclusivity
- *Trade secrets.* These include know-how and confidential information. Maintenance of trade secrets can be important to engineering disciplines where there is significant know-how, for example, in manufacturing processes.
- *Copyright.* This protects a wide range of content and materials relevant to engineers such as documents, design drawings, and software.
- *Design rights.* These protect the appearance of a product.
- *Database rights.* These protect collections of data and are important where there has been significant investment made in collating data that is to be made publicly available.
- *Trademarks.* These can be protected through registration or through actions for passing off. They are particularly important for engineering businesses where the value of the brand may lie in a reputation for safety or innovation.

The whole subject of intellectual property is quite specialised. Engineers working in fields that develop new IP need to be aware of the principles of IP protection but should work with IP lawyers to put appropriate protection in place.

The cost of patent protection can be quite significant for small companies, running into several tens of thousands of pounds or dollars per year for worthwhile protection of new technologies. This often leads to selective drafting of patents covering the really significant core areas of the technology rather than taking a broad-brush approach. There are some (large) companies who patent very little and rely on speed to market and the ability to stay ahead of the competition as an alternative to formal IP protection. They also argue that patent applications can alert the market to a general idea that competitors then try to develop themselves but working around the exact details of the patent.

A further issue can arise in certain fields, such as biotechnology and telecommunications, where there is such a high level of existing IP protection that developers of new ideas have to hack their way through a 'patent thicket' – see

Ref. 6.10 – incurring significant time and cost in the process. This has led to debate as to whether international patent legislation strikes the right balance between protecting inventors and stifling innovation. An IP assessment, and ideally IP protection, may be necessary to obtain external funding for an innovation.

Newton once wrote: 'if I have seen further it is by standing on the shoulders of giants'. Most engineering innovation is an extension of what has gone on before, relying on several centuries of intellectual property, from which the real innovation must be distinguished.

6.16 Funding of Early-Stage Work

Funding of early-stage work is a subject in its own right and one that is quite complex. Reference has already been made in Chapter 3 to the so-called 'valley of death', referring to the difficulty of moving from a demonstrated concept to the point where sales revenue is being generated. Chapter 10 covers this topic in more detail, particularly the different sources of potential funding and where they might come into play.

6.17 Concluding Points

The concept phase of development is the period when new ideas are put forward and fleshed out to understand whether they have merit as marketable, and financially viable, propositions. The activity will be technically led but should be conducted in a rounded manner, taking full account of wider business factors. Subsequent phases of work could be a lot more expensive and may build up some customer expectations, so it is important that the right concept is selected and developed. The work is best conducted in a multifunctional team environment where numbers of people are kept quite small – the work is not easily subdivided and is fast changing, so keeping a larger team coordinated can be difficult. Formal documentation of the work is helpful as a means of capturing what has been done and as a discipline to ensure that the concept has been fully thought through with no inconsistencies.

References

This article provides useful insights into successful practices with early-stage development:

6.1 Koen, P.A., Bertels, H.M.J., and Kleinschmidt, E. (2013). Effective practices in the front end of innovation. In: *Essay in the PDMA Handbook of New Product Development*, 3e. Hoboken, NJ: Wiley.

The concept of having (in a large company) a portfolio of new ideas is covered in this HBR article:

6.2 Nagji, B. and Tuff, G. (2012). Managing Your Innovation Portfolio. *Harvard Business Review* .

This book introduced the concept of 'open innovation':

6.3 Farrukh, C.J.P., Phaal, R., and Probert, D. (2010). *Roadmapping for Strategy and Innovation: Aligning Technology and Markets in a Dynamic World*. University of Cambridge, Institute for Manufacturing.

There are many books and articles on technology roadmapping. This one presents a good overview and is published by an organisation with plenty of experience in the field:

6.4 Chesbrough, H.W. (2003). *Open Innovation: The New Imperative for Creating and Profiting from Technology*. Boston: Harvard Business School Press.

This book is a very comprehensive survey of practice, focussing on the automotive industry:

6.5 Clark, K.B. and Fujimoto, T. (1991). *Product Development Performance*. Boston: HBS Press.

Published some time ago, Taguchi's principles are described in this publication:

6.6 Taguchi, G. (1986). *Introduction to Quality Engineering*. Asian Productivity Organization.

Pricing of innovations is an important but not a widely discussed topic, and these two references provide some useful insights:

6.7 (2014). *Global Pricing Study*. Simon-Kucher & Partners.
6.8 Ramanujam, M. and Tacke, G. (2016). *Monetizing Innovation*. Hoboken, NJ: Wiley.

The first publication is a good, short introduction to intellectual property whilst the second deals with some of the issues with IP protection:

6.9 *Intellectual Property Guide for Engineers* - Bird & Bird LLP, in collaboration with The Institution of Mechanical Engineers, 2015
6.10 Navigating the Patent Thicket, Cross-Licenses, Patent Pools and Standard Setting – Carl Shapiro, University of California at Berkeley, 2000. ISBN 0–262-60041-2

7

Identifying and Managing Engineering Risks

7.1 Introduction

Chapters 2 and 6 drew attention to two fundamental, and sometimes opposing, aspects of product and technology development work:

- The creative element of the process, developing new ideas to solve problems and to improve people's well-being.
- The element of risk which those new engineering solutions might introduce in terms of reliability, robustness, or creation of other forms of harm or danger.

This chapter is concerned with the second of these points – the identification of risks that require attention. Chapter 8 then deals with how those risks might be reduced or eliminated through engineering development work.

Risk is a widely used term, and its simplest definition is perhaps given by the *Cambridge English Dictionary*: 'the possibility of something bad happening'. Within the engineering and technology community, it is essentially 'the possibility of something going wrong'. A more technical description can be found in the *Risk Management Guide for DoD Acquisition*: 'Risk is a measure of the potential inability to achieve overall program objectives within defined cost, schedule, and technical constraints'. Risk in this context is normally considered to have two elements: the frequency of a potentially hazardous event and the severity of its consequences.

It should be noted that the concern in this book is with engineering or technical risks that can be addressed through the engineering development process. Other forms of risk, such as business or financial risks, are not covered here, although the same principles can be used for their management – see Refs. 7.1–7.3.

Managing Technology and Product Development Programmes: A Framework for Success,
First Edition. Peter Flinn.
© 2019 John Wiley & Sons Ltd. Published 2019 by John Wiley & Sons Ltd.

7.2 Identification of Risks

It might be argued that the separate identification of risks is a relatively new concept in the engineering world. Engineers have, over a 200- to 250-year period, always performed experiments and calculations to ensure that their products will work. Initially, knowledge was limited and failure was commonplace despite engineers' best efforts. For example, nineteenth-century railways were plagued with problems and several royal commissions were set up in the United Kingdom looking, for example, at iron bridges on railways (1847). Unlike today, that was a period when failure in service was commonplace and was one of the primary learning mechanisms – an approach that would be unacceptable nowadays.

Explicitly listing out risks and following them through to close out is a more recent phenomenon. The 'failure modes and effects' (FMEA) methodology, for example, which is one of several ways of analysing risks, was developed in the 1950s. Along with other approaches, it is used as a structured method of analysing potential failures and their effect on overall system reliability. This approach started in the military and aerospace industries, spread to automotive, and is now used widely in a range of engineering sectors. See Refs. 7.4, 7.5 for some further material.

This attention to potential failure has been effective and has produced results. On a lighter note, consider the famous 1896 London to Brighton car run to celebrate the lifting in the United Kingdom of the 4 mph speed limit for 'light locomotives' (cars, in today's parlance). There were 58 entries, 32 or 33 made it to the start line and somewhere between 13 and 20 finished the approximately 60-mile journey. The doubt about the numbers revolves around what constitutes finishing – does arriving a couple of days later after a major rebuild count as having finished? Compare this with today's cars, 120 years later. Reliability is measured in 'failures per hundred vehicles' and the best achieve about 0.7 fphv over a full year. *Failures* here are defined as any problem requiring attention, of which breakdowns are a small proportion. Rough arithmetic suggests that reliability has improved by a factor of about 10^5 over this period.

On a more serious note, the other major driver of improvement has been the investigation of major accidents. Public investigations into disasters such as Flixborough (1974), Seveso (1976), Three Mile Island (1979), *Challenger* space shuttle (1986), Kings Cross fire (1987), and Piper Alpha oil rig (1988) have fundamentally changed the approach to safety engineering, particularly in high-hazard industries where there is the possibility of major societal losses as well as the losses to the operator.

Subsequent sections of this chapter explain some of the approaches that have made these improvements possible. The overriding dictum is that successful design should fully anticipate all the possible and relevant ways in which failure can occur.

7.3   Risk-Based Approach

The risk-based approach is a response to the need for improved reliability and safety driven by the following:

- In the consumer sector, reliability as a competitive advantage, automotive products being a specific example where Japanese manufacturers set new standards that others then had to meet
- In other, high-risk industrial sectors, the need for absolute safety – nuclear power, space exploration, high-speed rail, chemical and process industries, oil and gas exploration, and passenger aircraft being examples
- In the military field, the need for reliability of weapon systems (and obviously safety, in the case of nuclear weapons)
- Allied to these points, the increasing complexity of products and systems, plus their increasing level of automation and reliance on software-based control systems
- An increasing public unwillingness to tolerate failures and their consequences

Engineering arguments might also be added to the list above. For example, development of designs is increasingly reliant on different forms of computer simulation and modelling, in turn leading to reduced physical testing. There is less opportunity, then, for the application of practical engineering instincts that have, for decades or centuries, been the backstop for the application of common sense. Risk analysis is, in some ways, also a response to this situation.

A similar, risk-based approach is also widely used in a range of fields of human activity: financial, business, health and safety, medicine, and project management being just some examples. An international standard, ISO31000:2009, and the associated document ISO Guide 73:2009 (Refs. 7.1, 7.2), give very general guidance, which is intended to be applicable to 'any public, private or community enterprise'.

At the core of the risk-based approach, irrespective of the application, are three straightforward activities:

1) Identification of potential hazards or risks
2) Estimation of the likelihood of each of these
3) Estimation of the severity of the consequences of potential failures, should they occur

The first of these is often conducted as a group, in which brainstorming is used to identify as many potential risks as possible, using the collective experience of the group involved – bringing in as far as possible the lessons of the past. The second and third are more analytical and, in some instances, may be quantified in terms of probability of occurrence over a defined period of time and in terms of the monetary value or the injuries/loss of life as a consequence of the failure.

Likelihoods and consequences are often described qualitatively on a 1–5 scale, such as the example shown in Figure 7.1, which is taken from MIL-STD-1629 Rev. A. Other standards and guidance documents, which are numerous, have something similar. Some extend the scales to 10 steps or further.

At the simplest level, a straightforward multiplication of likelihood and consequences (Figure 7.2) is often used to give an initial prioritisation to identified potential problems. This specific model has just three categories: usually designated as red, amber, and green. In engineering practice, a somewhat more complex approach is often used; in fact, matrices up to 14 × 14 have been seen and with individual prioritisation instructions in each of the 196 cells, but this is perhaps a little too complicated and prescriptive.

It should be noted that all risks, once identified, must be actively managed and not just dismissed. However, it is acceptable to judge a risk to be too unlikely to warrant attention, or for the cost of dealing with a risk to be out of balance with the likelihood and consequences, but these must be conscious decisions. The latter can be controversial, when the safety of workers or the general public is involved, but there are guidelines for decisions of this type – see below.

Level	Likelihood, Probability of Occurrence	Severity of Consequences
1	Extremely unlikely	None
2	Remote	Minor
3	Occasional	Marginal
4	Reasonably probable	Critical
5	Frequent	Catastrophic

Figure 7.1 Likelihood and consequence categories.

RISK ASSESSMENT MATRIX

LIKELIHOOD						
	5	5	10	15	20	25
	4	4	8	12	16	20
	3	3	6	9	12	15
	2	2	4	6	8	10
	1	1	2	3	4	5
		1	2	3	4	5
		CONSEQUENCES				

Figure 7.2 Risk assessment matrix.

A further development of this approach is to introduce a third variable: the likelihood that a failure mode will be detected at the design stage by the designer or those overseeing the design, or at the manufacturing stage, or during the operational phase. These three parameters (likelihood, consequence severity, and detectability) are then ranked on a 1–10 scale and a 'risk priority number' (RPN) calculated by multiplying the three parameters, resulting in an RPN between 1 and 1000.

7.4 Sources of Engineering Risk

One of the early standards in this field, MIL-STD-1629, defines the mode of a failure, which is effectively a risk that has materialised, as:

> *the physical or chemical processes, design defects, quality defects, part misapplication, or other processes which are the basic reason for failure or which initiate the physical process by which deterioration proceeds to failure.*

For any engineering product, simple or complex, there are many potential forms of failure, although their root causes may derive from a relatively small number of basic causes such as:

- Design-related:
  - Failure of the design to achieve the specified performance
  - Progressive loss of performance over the life of a product
  - Failure of the design to comply with legislative requirements
- Manufacturing-related:
  - Failure due to out-of-specification manufacture
- Mechanical failures:
  - Breakage due to overload during operation
  - Breakage due to fatigue damage accumulated over a period of time
  - Failure due to creep
  - Wear out
  - Corrosion, erosion, UV, or chemical attack
  - Delamination of composite material
- Electronic component failure
- Software design malfunctions

Potential failures, or risks, may be described in terms which relate more closely to the product, e.g. brakes 'stick on'. However, the underlying cause might be one of the above, such as corrosion. It is the latter information that will point to a solution which, in this instance, might be some form of protective coating, lubrication, or maintenance action.

7.5 Qualitative Risk Assessment Methodologies

Several methodologies exist for identifying, prioritising, and managing risks. The emphasis for all of them is to provide structure to the risk management process so that nothing is missed and so that there is an audit trail through to close-out of potential problems. This information is also a valuable record of learning and experience for use on future projects and can be used as the basis of troubleshooting during service operation.

In product-based industries (as opposed to process industries), 'failure mode effects analysis' (FMEA), is the most widely used methodology. The name is sometimes extended to 'failure mode, effects and criticality analysis' (FMECA). Figure 7.3 shows a typical template for an FMEA analysis.

The worksheet can be used to illustrate the step-by-step process of an FMEA study:

- Description of the design function that is required
- Identification of the potential mode of failure
- Description of the likely effect of the failure
- Ranking of the severity of the effect of the failure on a scale of 1–5 or 1–10
- Description of the mechanism of the potential failure
- Ranking of the likelihood or probability of the failure, also on a 1–5 or 1–10 scale
- Description of the means by which a fault would be detected in service
- Ranking of the detectability on a 1–5 or 1–10 scale
- Calculation of a RPN, obtained by multiplying *severity* × *likelihood* × *detectability*, resulting in an RPN between 1 and 125 or 1 and 1000
- Agreement on the course of action to be taken
- Agreement on who will accept responsibility and the timescale for completion
- Summary of the action taken
- Revised severity/likelihood/detectability/RPN based on the action completed

A simplified version would omit the detectability ranking and just rely on severity and likelihood.

The FMEA process should be initiated as soon as schematic drawings or descriptions are available and revisited or updated periodically as a living document in much the same way as drawings or technical specifications. The later the start is made to an FMEA, the less will be its influence. Any modifications to a product should be evaluated through the FMEA process.

Early-stage FMEA analyses will point towards analysis, modelling, or test work that should be used to examine potential failures and reduce their likelihood. They may also be used as a method of evaluating alternative solutions.

FMEA WORKSHEET

Item / Function	Potential Failure Mode(s)	Potential Effect(s) of Failure	Severity	Potential Cause(s)/ Mechanism(s) of Failure	Probability	Current Design Controls	Detectability	Risk Priority Number	Recommended Action(s)	Responsibility & Target Completion Date	Actions Taken	Action Results			
												New Severity	New Probability	New Detectability	New RPN

Figure 7.3 Typical FMEA worksheet.

The FMEA method, as a ground-up approach, works well at the level of components or subassemblies and with single modes of failure. It is sometimes described as an 'inductive' approach and depends on experience of similar products and systems. For those designing and supplying these elements of a product or system, FMEA is often mandated by the system integrator, who may use a combination of individual FMEAs to identify critical characteristics and assess system reliability and failure consequences.

The same approach may also be used in a manufacturing environment to analyse potential problems with manufacturing processes – so-called process FMEAs, or PFMEAs. The structure and approach follow the same principles, including the listing of critical characteristics at each step of the process and putting in place an appropriate control plan to make sure that the process is generating good product.

It should be noted, however, that FMEA is less effective with complex systems and with multiple, interacting modes of failure. For these more complex situations, other methodologies such as fault tree analysis (FTA) can be used in combination with individual FMEAs.

7.6   Fault Tree Analysis

Whereas FMEA is a bottom-up approach, FTA looks at complex systems from the top of a system downwards.

The method was developed initially within the US military in the 1960s as a way of improving the reliability of missile systems. Its use since then has extended into the nuclear power, oil and gas, chemicals, aerospace, rail, and automotive sectors – all applications involving complex and critical systems.

The starting point is the definition of a significant system failure, such as failure of an engine to start, often referred to as an 'undesired event'. From there, potential causes of failure, located more deeply in the system, are analysed and documented – the potential sources of failure, the causes of the causes, and so on, in the form of a hierarchy or tree of events. Thus, in the example suggested above, the 'failure of engine to start event' might be broken down thus, very simplistically:

- Lack of fuel
  - Pipe blockage
  - Pump failure
  - Pipe disconnected
  - Run out of fuel
- Failure of starter motor to operate
  - Battery failure
  - Motor jammed

- – Motor open circuit
- – Cable disconnected
- – Electrical supply failure
- Failure of ignition system
  - – Damaged/dirty plugs
  - – Leads disconnected
  - – Coil failure
  - – Electrical supply failure

Figure 7.4 shows diagrammatically the layout of a typical fault tree.

As can be seen, a complete tree covers the top event down to basic causes – which might already have been identified in an FMEA-type of study. Numerical probability failures may be estimated for the basic events from which overall system reliability can be calculated. Software is available to support the analysis of complex systems.

The fault tree is essentially built around Boolean logic, and there is a standard notation for such trees with agreed symbols. Examples of the latter include: basic (bottom-level) events, 'AND gates', where two failures have to combine to produce the higher event, and 'OR gates', where either of two events can cause the higher event.

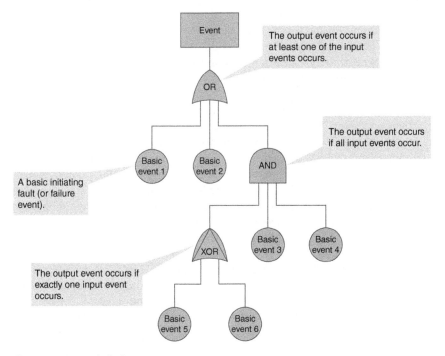

Figure 7.4 Example fault tree.

The approach is good at exploring complex systems and multiple sources of potential failure. It can look at the effectiveness of redundancy in a system, the interaction of failures, and identify common causes (one fault, such as loss of electrical power, leading to multiple failures). Its basic purpose is to draw attention to critical risks, which will then be the subject of redesign or development effort either to eliminate them, or to reduce their impact, or to include them in system monitoring. The approach is also relevant to maintenance work and fault diagnosis

7.7 Hazard and Operability Reviews – HAZOP

The examples described above relate to both products, such as aircraft or rail vehicles, and to fixed, safety-critical plants. A related approach, now widely used, comes originally from the heavy chemicals sector but is applied to other forms of fixed plant as well as, occasionally, products such as rail vehicles. Known universally as HAZOP, the full title is 'hazard and operability' studies and the emphasis, as the name implies, is on issues that might arise during operation – see Refs. 7.6–7.9.

The method breaks the complex plant, process, or asset into a series of 'nodes', each of which is examined systematically in detail, again using a team-based, multidisciplinary approach in a series of workshops spread over a period of time. Each node is examined for potential deviations from normal operation, which could cause hazards and operability problems. There is a lexicon of standard keywords which is used to prompt the team to identify issues. These are generally very simple words such as: *more, less,* and *late.* An experienced scribe is usually appointed to the team and a typical worksheet is shown on Figure 7.5. As this worksheet shows, the basic elements of the methodology is very similar in principle to FMEA, with its concentration on identifying potential problems and then ensuring that they are managed.

Whilst this method can be applied retrospectively to existing plants, its real value comes from application during detailed design; a HAZOP study can be run once there is process definition and piping and instrumentation diagrams (P&IDs). The method is effectively mandated in many situations where regulatory approval will only be given if HAZOP studies, amongst other things, have been completed. It is, however, a time-consuming and expensive process and there may be a temptation to restrict the scope of a HAZOP analysis to safety and environmental concerns thus excluding reliability and efficient plant operation. Its application is most relevant at the detailed design stage. Similar but somewhat simpler methods, such as hazard identification studies (HAZID), can and should be applied at earlier stages – concept design and 'front-end engineering design' (FEED) stages.

HAZOP STUDY REPORT

DATE:
NODE:
DESIGN INTENT OF THE SYSTEM:
HAZOP TEAM CHAIRMAN:

P & ID NUMBER:

HAZOP TEAM MEMBERS

1)
2)
3)
4)

Item	Process Parameters	Guide Word	Deviation	Possible Cause	Possible Consequences	Existing Safeguards	Action Required	Action by	Action taken

Figure 7.5 Typical HAZOP worksheet.

7.8 Quantitative Risk Assessment

The methods described above are essentially qualitative although some, such as FTA, can be used either qualitatively or quantitatively. Quantitative risk assessments (QRAs), or probabilistic risk assessments (PRAs), are normal practice in certain complex, high-hazard industries, such as oil and gas, chemicals, and nuclear power generation. Their origins are again in these fixed, one-off plants or installations, rather than volume products. However, similar quantitative methods are growing in use in the latter field and arguably have been used for many years in aerospace and rail, albeit in different forms from the process industries and usually described as 'reliability studies'.

The start point for quantitative methods is the same risk identification approach described above, listing potential risks and assigning likelihoods and consequences in a qualitative manner. The more serious risks are then analysed quantitatively, the objective being to establish numerically the likelihood of major incidents and the consequences, again numerically, of such incidents. This then permits an evaluation of the figures against those that are considered acceptable to society (which is a subject in its own right).

Likelihoods are usually expressed as the probability of occurrence over a defined period, such as a year, or the probability per mission – a typical figure might be 1 in 10^5 or 0.001%. Consequences could be measured in terms of injuries, deaths, financial cost, environmental damage, or loss of aircraft. Various modelling techniques are used to evaluate consequences, e.g. blast damage, atmospheric dispersion, radiation, toxicity, and environmental pollution.

The methods for analysis of this type are essentially empirical and have been built up by experience over a number of years, on an industry-by-industry basis. Unlike other forms of engineering modelling, simulation methods cannot be verified by running experiments and correlating results. Accidents and incidents are studied avidly, but there is an inherent level of uncertainty in most quantitative safety analyses that then leads, rightly, to a cautious approach being taken by the application of a 'disproportion factor', a form of safety factor, to the analysis.

The sectors where quantitative safety analysis is performed are generally those where regulatory approval is based on a safety case (see below). The quantitative analysis is therefore one of several means of justifying that approval and not an activity conducted in isolation.

7.9 Functional Safety

An increasing number, if not the majority, of products, machines, plant, and equipment rely heavily on electronic control systems and software for their normal functioning. At one extreme, some systems are wholly reliant on active, automated control for a normal operation; for example, some military aircraft are aerodynamically unstable (to increase their manoeuvrability) and are hence completely reliant on automated control systems for safe flight. Similarly, many complex plants are beyond the stage where they can readily be controlled manually. Looking ahead, there is much interest (e.g. in cars and ships) in developments towards autonomous operation. In all these situations, safety is dependent on the components, equipment, and control systems working correctly.

'Functional safety is the term used internationally to describe the situation where safety has this dependence on correct operation. A series of international standards cover the topic, with IEC61508 as the master standard. There are then separate and specific standards for automotive, rail, process industries, and nuclear – see References.

Within these standards, a 'safety function' is a process in which an active control system detects the development of potentially dangerous conditions and triggers protective or corrective actions, either as a separate override system or as part of the wider functionality of the core system.

As indicated above, the topic is relevant to a wide range of industries: process plants, oil and gas, nuclear, medical, rail signalling, automotive (especially as autonomous technologies develop), aerospace, and machining. Even simple devices, such as automatic door opening systems in shopping malls, have a degree of safety criticality.

As with all potential hazards, the starting point for determining what should be done to address the safety of such systems is an initial risk analysis. This may or may not show the need for active functional safety (other forms of safety, such as mechanically interlocked guarding, may be sufficient). The IEC61508 standard (Refs. 7.10, 7.11) accepts that zero risk cannot be achieved, but it does promote the consideration of safety from the outset of system design and make the point that nontolerable risks must be reduced to as low a level as reasonably practicable (see Section 7.10). The standard is 'end-to-end' in the sense that it applies to all points in the life cycle from concept to disposal.

The International Electrotechnical Commission (IEC) standard uses hazard and risk analyses based on six categories of frequency and four categories of consequence severity, giving a 24-cell risk assessment matrix. From this matrix, four categories of risk are then identified:

- *Class I*. Unacceptable in any circumstance
- *Class II*. Undesirable: tolerable only if risk reduction is impracticable or if the costs are grossly disproportionate to the improvement gained
- *Class III*. Tolerable if the cost of risk reduction would exceed the improvement
- *Class IV*. Acceptable as it stands, though it may need to be monitored

SIL LEVEL	Probability of failure on demand (i)	Probability of failure per hour (ii)
1	$10^{-1} - 10^{-2}$	$10^{-5} - 10^{-6}$
2	$10^{-2} - 10^{-3}$	$10^{-6} - 10^{-7}$
3	$10^{-3} - 10^{-4}$	$10^{-7} - 10^{-8}$
4	$10^{-4} - 10^{-5}$	$10^{-8} - 10^{-9}$

Figure 7.6 Safety Integrity levels.

These categories can then be used to decide whether a safety function is needed to achieve a sufficiently low overall risk. If a safety function is needed, it must then be decided what form it should take in order to achieve and maintain a 'safe state' within the system being controlled (which could include shutdown) in the event of problems arising. The performance criteria for this safety function are in turn defined in terms of safety integrity levels (SILs) (Ref. 7.12), which specify probabilistically the likelihood of not achieving the defined function under defined conditions within the required timescale – see Figure 7.6. These levels in effect specify the amount of risk reduction that the safety function must provide relative to an unprotected situation.

Two types of system are considered: those ('i' above) that switch in when a problem is detected (e.g. an emergency shut-down system) and those ('ii' above) that operate continuously.

The material above is intended as a very broad summary of the concept of functional safety and how it should be assured. There are significant differences how different industries address the topic, reflecting their different needs and the different consequences of failure. For example, machine tools have much less capacity to cause large-scale damage and injury than a major process plant. There is also no uniform definition of safety integrity levels (SILs) across all industries. However, the risk-based approach and the concepts of functional safety are widely accepted and will apply to a wider range of engineering products and systems as they adopt an increasing level of built-in 'intelligence'.

7.10 As Low as Reasonably Practicable

The question was raised earlier as to what level of risk is considered acceptable when the safety of users, operators, employees, or the general public is concerned. The UK health and safety regime operates under the legal umbrella of 'as low as reasonably practicable', abbreviated to the ugly but widely used acronym of 'ALARP'. This approach developed from workplace safety considerations but the principle can be applied more widely to hazardous plants and to other safety or risk-related situations – see Ref. 7.13.

The approach recognises that absolute safety cannot be achieved, and it is acceptable (and legally defensible) to choose not to adopt certain safety measures. Thus, if certain measures are not technically feasible (i.e. not practicable) or are not proportionate to the benefits (i.e. not reasonably practicable), then they can be rejected.

However, the test of what is reasonable can be strict. It must be shown that existing good practice, as it exists in other similar or related situations, has been adopted. To support this argument, it would be normal to carry out a full qualitative risk assessment and probably a quantitative assessment. The latter, in particular, would show which safety improvement measures should be adopted and which do not bring a proportionate benefit. It should also be noted that the state-of-the-art is always moving forward: new technologies are being developed and learning will be acquired from accidents or incidents. Hence, what might be considered acceptable in one decade may not be acceptable in the next.

The ALARP concept is applicable within certain boundaries – see Figure 7.7, where the width of the inverted triangle represents the frequency of the risk. If, for example, after adopting all reasonably practicable measures, the level of quantified risk for a product, facility, or activity is still above certain boundaries,

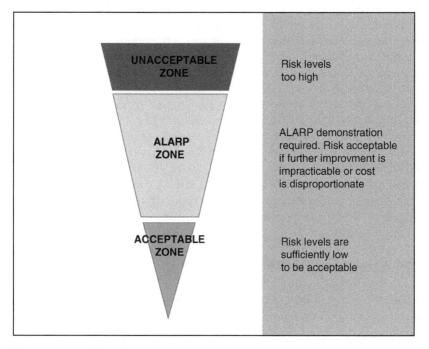

Figure 7.7 ALARP boundaries.

then the proposal will have to be abandoned as unacceptable or an alternative solution developed. Conversely, once risk reduction measures have reduced risks to negligible levels, then the ALARP principle no longer applies. In this context, the ALARP zone is considered to be in the approximate range of 1 in 10^3 to 1 in 10^6 risk of a fatality in one year.

7.11 Safety Cases

The points above relate mainly to single risks or failures, each of which is considered in isolation. With every product, system, or plant, multiple risks will accumulate to an overall level of risk, which may or may not be acceptable. For complex systems or plants, it is normal to compile a 'safety case' that examines all aspects of risk and puts forward arguments as to why the system overall is adequately safe. Such a safety case is required for regulatory approval in a growing range of industries: nuclear, oil and gas, chemicals, rail, medicine, aerospace, and automotive as examples.

In this situation, a safety case is a compilation of the arguments that the system in question is adequately safe when operating for a specific purpose under specified operating conditions. This is sometimes described as an evidence-based approach where the evidence supports the arguments that the system is safe. The alternative is a more prescriptive approach where, by complying with certain standards (national or international) or test methods, something is considered safe. The latter approach is perfectly acceptable for small-scale products, such as small domestic appliances, where international standards or test codes are the normal means by which safety is assured. The safety case approach is applicable to more complex systems where there are multiple potential failure modes and the possibility of failures interacting with each other. The case considers not just the design of the system but how it is operated and maintained, how staff are trained, what emergency procedures exist, and how these factors are all managed.

Because of the arguments-based nature of safety cases, and hence the potential for different interpretations, regulatory approval may require an independent peer review before approval is given.

7.12 Stretching the Boundaries

Chapter 2 introduced the concept of risk in engineering and identified four categories of risk, which are repeated below:

1) *Obvious.* Those that both the developer of the technology and outside parties agree upon and where therefore consensus can be easily achieved

2) *Experience.* Those that the developer may have missed but that peers and grey-haired engineers might identify from their experience
3) *Hidden.* Those for which the signs are present but the developer might be ignoring or dismissing, e.g. because test material is 'nonrepresentative'
4) *Danger zone.* Those that lie beyond the experience or expectation of all concerned

The approaches described so far should bring out risks in the first three of these categories, provided they are done thoroughly and involve experienced people. But this then raises the question: how can risks that are beyond normal experience be identified? Clearly, the more thorough and rigorous the risk analysis, the smaller is the danger zone. But is there any guidance that could be suggested to identify the 'unknown unknowns'? Logically, if they are unknowable, then they are destined to be hidden until they emerge, taking everyone by surprise. However, there is merit in any risk review in stepping back at some point and asking: 'what have we missed?' Areas to consider might include:

- Taking an established technology but stretching its application just too far
- Subjecting an established product to a duty cycle with which it cannot cope
- Not foreseeing potential software and control system problems
- Malevolent hacking of control and communication systems
- Use and abuse of products outside the operating envelope originally envisaged

The book *Design Paradigms* by Henry Petroski (Ref. 7.14) contains some instructive examples, over a long period of time, from the field of bridge-building. It is interesting to see how particular forms of bridge design move from conservatism, in the early stages, to overconfidence and failure, then back to conservatism. For example, the Tacoma Narrows suspension bridge, which broke up in high winds in 1940 as a result of aerodynamics forces, is often regarded as the revelatory event for this phenomenon. However, history shows that at least 10 suspension bridges were destroyed or severely damaged by wind in the nineteenth century alone. Another instructive book is *Major Hazards and Their Management* (Ref. 7.15).

In the twenty-first century, it could be argued that software, control, and communication systems present the greatest challenges in terms of stepping into uncharted territory. The technologies of the Fourth Industrial Revolution undoubtedly present tremendous opportunities, but their downsides and risks need also to be given careful thought.

Hence, good practice in risk management should always include a brain-wracking activity to try to identify such issues. At the same time, care must be taken to ensure that the culture of the organisation concerned allows people to raise concerns without fear of ridicule or retribution.

7.13   Concluding Points

Identifying, reducing, and managing technical risks is one of the most fundamental aspects of the engineering development process. It has been an integral aspect of this process from time immemorial, but it has been an explicitly managed topic since around the 1960s. Partly driven by the increasing complexity of products and systems and partly by greater market pressure for reliability and safety, it is a topic in its own right with an extensive literature which this chapter can summarise only briefly.

Risk identification and management is one of the primary mechanisms for embodying the lessons of the past. As George Santayana said, 'Those who cannot remember the past are condemned to repeat it' and this is particularly true of engineering work. Learning from the failures of the past is central to the engineering process. It might also be considered legally negligent not to do so.

Much of the basic thinking and methods of analysis in this field has come from two sources:

- *Aerospace and defence.* Driven by the needs of reliability and, in the case of passenger aircraft, basic safety
- *High-hazard process industries.* Driven by the potential consequences of major accidents

The basic methodologies of risk management all revolve around the simple concepts of identifying potential hazards, estimating their likely frequency and consequences, and then adopting courses of action appropriate to the level of risk. It is accepted that zero risk is impossible to achieve and problems or accidents do still happen. However, all long-term measures of reliability and safety show a strongly improving trend and what was considered acceptable some decades ago would be intolerable today. This trend will continue into the future.

At the same time, the future will present challenges. In particular, most products, and not just complicated items such as aircraft or major plants, are incorporating an increasing level of intelligence. Predicting how they will operate and what could go wrong will stretch the minds of both engineers and regulators.

For engineers wishing to educate themselves more widely on this important topic, formal investigations into major accidents or problems (which go back to the 1840s) are an interesting source of learning for the enquiring mind. The older case studies have the advantage of not being clouded by political or social factors, litigation, or vested interests. Although the technologies involved have been superseded, the general points of learning are still valid, especially as most of them revolve around human factors and human error. These case studies are concentrated in industries that have can have a wide societal impact if something goes wrong, but the lessons of the past apply to all fields of engineering.

References

International standards and public procurement documents provide overall frameworks for risk management:

7.1 ISO31000:2009 – Risk Management, International Standards Organisation, 2009
7.2 ISO Guide 73:2009 – Risk Management Vocabulary
7.3 Risk Management Guide for DOD Acquisition, 6, (Version 1.0)

FMEA methods are covered by a number of documents including:

7.4 MIL-STD- 1629 Rev. A, US Department of Defense, November 1980
7.5 SAE J1739_200901 (Revision 4) - Potential Failure Mode and Effects Analysis in Design (Design FMEA), Potential Failure Mode and Effects Analysis in Manufacturing and Assembly Processes (Process FMEA)

HAZOP methods are covered in these four references:

7.6 Hazard and Operability Studies (HAZOP Studies) Application Guide BS: IEC61882: 2002
7.7 Kletz, T. (2006). *HAZOP and Hazan*, 4e. Taylor & Francis.
7.8 (2008). *HAZOP: Guide to Best Practice*, 2e. European Process Safety Centre, IChemE.
7.9 IEC 61882:2016 (2016). *Hazard and Operability Studies (HAZOP Studies) - Application Guide*. International Electrotechnical Commission.

Functional safety and safety integrity levels are covered in IEC61508 and related standards:

7.10 IEC 61508 (2016). *Functional Safety of Electrical/Electronic/Programmable Electronic Safety-Related Systems (E/E/PE, or E/E/PES)*. International Electrotechnical Commission.
7.11 Related standards: IEC61511 (process industries), IEC 61513 (nuclear power plants), IEC 62061 (machinery systems), IEC 62425 (railway signalling systems), and ISO 26262 (road vehicles).
7.12 Health and Safety Executive (2004). *A Methodology for the Assignment of Safety Integrity Levels (SILs) to Safety-Related Control Functions Implemented by Safety-Related Electrical, Electronic and Programmable Electronic Control Systems of Machines*. UK.

This paper gives a good overview of the application of ALARP principles:

7.13 Back to Basics: Risk Matrices and ALARP, R. David & G. Wilkinson, 2009

Finally, these books are well worth reading:

7.14 Petroski, H. (1994). *Design Paradigms, Case Studies of Error and Judgement in Engineering*. Cambridge University Press.

7.15 Wells, G. (1997). *Major Hazards and Their Management*. IChemE.

8

Validation by Modelling and Physical Testing

8.1 Introduction

Chapter 6 described how the concepts for new products are formulated and Chapter 7 spoke about identifying the engineering risks that might need to be overcome for those products to be robust offerings to the marketplace. This chapter is concerned with the analysis, modelling, and test work, which could be used to identify and overcome those risks before new products or services are launched. The aim, as always, is to maximise the reliability of new products at their points of launch.

The approach is sometimes described in a V-shaped or 'waterfall' model, as indicated below (see Figure 8.1):

The left leg of the V signifies the design phase of the development, starting with the overall product and progressing through systems and assemblies down to the individual details. Then, the right leg of the V signifies the development and validation work, which is undertaken to identify any issues or problems that the design might possess, starting at the level of the individual component and working up to the complete system.

This is a good overview of what happens, and this approach is also used in other fields such as software development, but it does simplify how validation takes place, making it seem more of a neat, serial activity than is actually the case. Rather than just being a 'box-ticking' activity, it is, in practice, more of an iterative development process. However, it is the case that validation-type activities do take place at the level of the individual component, the system, and the complete product. It highlights that components must work individually but also must work together as a complete system.

The different forms of validation are described below, after confirming the overriding purpose of the work.

Managing Technology and Product Development Programmes: A Framework for Success, First Edition. Peter Flinn.
© 2019 John Wiley & Sons Ltd. Published 2019 by John Wiley & Sons Ltd.

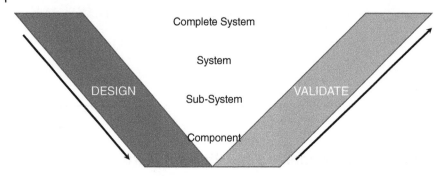

Figure 8.1 Waterfall development model.

8.2 Purpose of Development and Validation Work

In simple terms, the purpose of development and validation is to confirm that the product performs as intended and will continue to do so reliably throughout its working life. In summary, validation will cover:

- Performance testing to verify that the product or system functions as intended and delivers to the expectations of customers
- Testing to confirm that it will meet any relevant legislative, safety, or regulatory requirements, including formal acceptance tests
- Life testing to confirm that the product will continue to function to its requirements throughout its life and that the life is as expected
- Extreme testing, to confirm that the product will continue to function at the extremes of its operating envelope or when subject to abuse, up to a pre-determined level, or when operating in different environmental conditions
- Reliability testing to prove that the product will function dependably, in terms of failure rate or other similar measurements, throughout its design life
- Confirmation testing to demonstrate that the product made by volume production methods performs in the same way as those made by earlier methods

The validation activity, whatever form it takes, is intended to identify problems where the product does not conform to its requirements, understand their causes, and then test solutions to them. Many of the requirements may take the form of standards, either from public sources or applying within the organisation, in the case of larger companies. There should also be a particular focus on the risks identified earlier, with the intention of ensuring that they do not materialise as real-life problems. Both the standards and the identified risks bring, in effect, experiences of the past into the present. The intention is to ensure that any problems or defects are not allowed to escape and be detected by the customer.

The discipline can also be described as 'reliability engineering', and this title has some validity in the sense that the overriding goal is to achieve a reliable product. For the purposes of this book, however, a narrower understanding is taken of 'reliability'; as indicated above, if the term is used to describe the numerical failure rate (actually, one minus the failure rate see Section 8.13) associated with the product or system.

8.3   Methods

For the purposes of discussion in this book, methods of development and validation are broken down into three distinct groups of activity:

- Calculation
- Modelling and simulation
- Physical testing

These groups are not mutually exclusive; many components or systems will be the subject of all three types of development, and in the order suggested. This is also the approximate order of increasing cost, so getting things right at the early stages will save money later, especially if repeat test work can be avoided. Experience accumulated over recent decades has shown that better modelling and simulation techniques have reduced substantially the need for later physical testing, although many organisations still want the confidence of physical sign-off, as is the case with many regulatory requirements. Modelling has also made possible solutions where trial-and-error physical testing is just not possible, space travel, and nuclear power being two examples.

8.4   Validation and Test Programmes

Stating the obvious, an integrated programme of development, validation, and testing should be drawn up as early as possible in a technology or product programme. This should be based on established practices and previous company learning but also covering the points raised in the risk analysis, some of which might be outside the organisation's earlier experience or be particularly critical to the success of the product. The programme will require several iterations and will only be capable of being drawn up in full detail when the technology has matured to TRL 5 or 6.

Validation work of this type is time consuming and often needs specialised skills and resources such as computing power and test facilities. If third-party facilities are needed, booking well ahead may be needed. Test items will also need to be made and quantities estimated. In the case of both resources and material, allowance should be made for failures and repeat tests, noting that the purpose of much testing is to cause failures and understand failure mechanisms rather than just run through a programme unscathed. One of the key aims of the programme should be to show up failure mechanisms as early as possible.

8.5 Engineering Calculation

The use of fundamental engineering theory is the most basic means by which the behaviour of new products can be predicted and evaluated prior to their physical realisation. Development of new ideas, for example using sketches and simple drawings, goes hand-in-hand with basic engineering calculations (see Figure 8.2).

The methods available are too numerous to list; they cover all areas of engineering from structures, materials, mechanics, fluid flow, and thermodynamics to electronics, turbo-machinery and nuclear engineering. The personal computer, and associated software, has revolutionised this unseen aspect of engineering by automating, and improving the accuracy, of much of the hard work that was formerly the province of the slide rule and data sheets.

Whilst spreadsheets are the obvious starting point for engineering analysis, most of the basic engineering theory is available, in effect pre-programmed, in

Figure 8.2 Engineering calculation – Sir Frank Whittle at work - before the days of computers!

proprietary software packages. Spreadsheets are good for purely numerical calculations, but are not as well suited to calculus and differential equations. Individual companies then have their own privately developed versions to support, confidentially, their own specialised in-house needs. However, proprietary software is no guarantee of accuracy, even if it is based on fully accepted theory. Correlation of new methods with real life is vital, and checking of calculations by more experienced staff, using their experience and common sense, is still an important part of the engineering quality system.

8.6   Modelling and Simulation

Computer-based modelling and simulation tools are now in ubiquitous use across the spectrum of engineering disciplines. The boundary between the calculation methods noted above and modelling is a matter of debate. Models are generally used to take analysis to a higher level of complexity and detail, examining complete products, systems, processes, and their performance. Conversely, calculation methods, as defined in this book, tend to have narrower boundaries and can be done relatively quickly. The latter is therefore more suited to the earlier stages of engineering development.

As an aside, it should be noted, however, that before the advent of computers, some very complex analyses were actually undertaken by hand. Nevil Shute's autobiographical book *Slide Rule* (Ref. 8.1) describes how the statically indeterminate R100 airship structure was analysed in the 1920s by a team of 'calculators' (Shute was 'chief calculator') over many weeks and with some errors that required backtracking. A finite element structural model would do the job today in a matter of minutes. The 2016 film *Hidden Figures* tells a similar story about calculation methods in the early days of NASA.

Models do take some investment to set up and depend on some level of engineering detail being available for the product or system being modelled. They are usually constructed by building up the model from elements representing subsystems or components of the product. Once in place, however, models can be run to produce simulations (*simulation* is the term used to describe the running of mathematical models) of a wide range of scenarios or conditions to explore a system's behaviour. This may include simulations of dangerous conditions or may replace physical tests that could consume a lot of expensive material, e.g. crash testing. Once established, they can be updated and re-run as a product develops, and their accuracy can be improved as physical data become available. Some software includes the ability to perform re-runs automatically and for the software to home in on the optimal solution.

Within the field of engineering, models are widely used. Examples of topics include engineering structures (see Figure 8.3), fluid flow, dynamic behaviour, electronic circuit performance, and atmospheric dispersion of gases. Outside

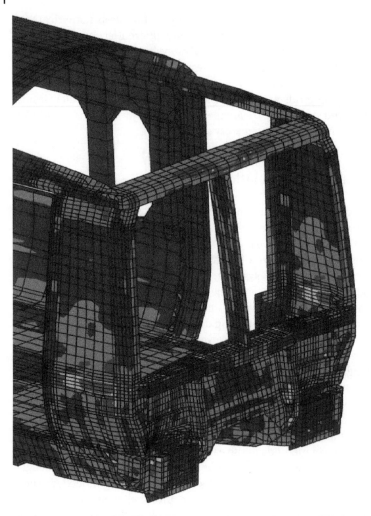

Figure 8.3 Example of finite element model of part of Rail Carbody Structure.

engineering, models are available for such diverse topics as weather forecasting, the spread of viruses, the first milliseconds of the universe, social behaviour, and traffic flow!

A distinction is usually drawn between models that are essentially static, representing one set of conditions, and those that are dynamic, tracing behaviour over a period of time. A static structural model would be an example of the first and a weather forecasting model an example of the second. The latter generally requires much more computing power, and a further distinction is drawn between those that move forward in discrete time steps versus those that produce a time-continuous output.

The growth of modelling and simulation has been facilitated by the reducing cost, and increasing availability, of computing power. This, in turn, has been supported by software developments that provide the mathematical models themselves around which the software operates. Often, this software also automates the initial compilation of models and provides results in impressive, easily digestible, and graphic form.

This then introduces one of the downsides of readily available modelling software: how accurate are they? The adage 'garbage in, garbage out' is very relevant. It could be argued that models are only as good as the correlation that has been made with the physical world. Where this correlation is in place, which is the case with well-established applications, models can go a long way towards replacing physical testing, as well as being able to model situations where physical testing would be hazardous, impossible, or just too expensive. However, a word of caution is appropriate: checking of simulations, both the models and the results, by more experienced staff is another important part of the engineering quality system.

A further extension of modelling, which some would consider a separate discipline, is 'virtual reality'. The ability to 'see' a product or system in three dimensions opens up the possibility of trialling its use and exploring its serviceability. Again, problems can be discovered at an earlier stage of development ahead of expensive hardware commitments.

The field of modelling is always moving forward with developments into increasingly difficult and complex areas, often requiring the use of high-powered supercomputers. For most engineering applications, however, tried and tested models and software are readily available.

The concepts described above relate mainly to prelaunch engineering development aimed at ensuring the integrity of the final product. The use of modelling and simulation techniques is being extended into the post-delivery phase by creating what are described as 'digital twins'. These models are constantly used, updated, and refined based on in-service data, supporting the operation of the product in the field and providing feedback for use in future design work – practical examples of the use of Internet of Things and Big Data technologies. The same approach can be taken in creating digital twins of manufacturing systems.

8.7 Physical Testing

Trialling of a new technology or product in physical form is the ultimate test of the idea and provides the closest match with the real-world operation. A trial programme could include testing by customers, in cooperation with the developer, in normal operating conditions.

A complete prototype of a complicated product may also be the first opportunity to test all the elements of a product together to reveal any complex and detrimental interactions between the systems making up the product. These interactions are difficult to model either by calculation or simulation. In practice, test work can be carried out on components, systems, or complete products, and may be conducted using laboratory rigs or under field conditions.

Test work can aim to prove a number of aspects of a new design, and different approaches are required for each. In principle, testing can examine:

- Functional performance
- Life
- Reliability
- Environmental performance
- Serviceability

Performance testing has the aim of proving that the design provides the benefits expected by, or specified to, the end customer. Instrumenting test pieces, cycling them through the product's operating envelope, and analysing the data are the means in principle by which questions are answered. Established companies have batteries of test codes for doing this work, often developed over many years and, in effect, representing the accumulated experience of the organisation in making reliable and successful products. Compliance with legislation may also need to be demonstrated by testing to internationally agreed test codes. Such tests are normally witnessed by the certifying body.

In the early stages of development, performance testing might be confined to relatively 'normal' operating conditions, representative of a careful operator in a controlled environment. This may be sufficient to prove the concept. However, a proportion of products will be abused, overloaded, or operated in some way outside a normal duty cycle. This could also include operation in hostile environments which might include extremes of:

- Temperature
- Humidity
- Vibration and shock
- Electrical surges and electromagnetic pulses
- Dust, salt, and dirt

Alternatively, a potentially dangerous condition may arise by accident and the product or system is expected to deal with it. Testing in abnormal conditions should therefore be planned. This could include extremes of temperature (hot and cold), extremes of load, electrical extremes, edges of the flight envelope in the case of an aircraft, or combinations of conditions that may be unlikely but nonetheless could arise. The risk analysis should define these requirements.

A conscious choice can be made about the extent to which these extremes should be covered. Regulatory requirements specify what is required in many industries but, where this is not the case, it may be uncompetitive to provide too robust a product. Nonetheless, it would be naïve to think that customers or operators are always very careful, and they do have a habit of pushing products to their limits. Alternatively, difficult or dangerous conditions may arise by accident, and the product must have the capacity to deal with them. The latter is more of an issue for completely new technologies and companies; established industries usually have empirical rules that cover this issue.

Testing of this type can be quite close to being representative of real life. However, whilst prototype material may look exactly like the final, production version, methods of manufacture will almost certainly be different. This can impact the accuracy of results, and some repeat testing of early production material should be considered.

Life testing presents more problems. If a product is expected to last 10–30 years, how can this be compressed into a test programme lasting a few months or a few years? There are several ways of addressing this issue. The simplest way is some form of overload testing, where a product is put through a much more arduous duty cycle than it would experience in real life. Crude though this approach may sound, there is some science behind it in certain circumstances. For example, if the life of a product's structure is determined by fatigue behaviour, accelerated overload testing, in effect, misses out the nondamaging low-level loads and just concentrates on the high-level damaging loads. This is the thinking behind pavé testing of road vehicles, which was introduced as an empirically based accelerated test method but which now has some theory and correlation to support it. This type of test can accelerate life by factors between 50 and 100. Care must be exercised to ensure that overload testing does not create unrepresentative failure modes.

An alternative, used in the aerospace industry, is to have a long-running fatigue test using realistic, in-service loads but always keeping ahead of flying aircraft. With this type of test, the timescale acceleration comes from concentrating on take-off/de-pressurisation/pressurisation/landing cycles, missing out the steady-state conditions.

If the product has a service life that involves intermittent use, e.g. a domestic appliance, then test units could be run on a continuous basis to accumulate running hours to match the full life in a short period. In this situation, though, care must be taken to simulate start-up and shut-down, which may be the determining factors in setting a product's life. In a similar way, locks, mechanisms, hinges, catches, and doors can be tested cyclically with a simple test rig, which can produce 5000–10 000 cycles per day. Tests of this type are usually stopped periodically to examine parts for wear or damage.

A further approach, more suited to components rather than larger products, is accelerated testing in climatic chambers where heat, cold, humidity, corrosive effects, or dust may be simulated. Such chambers can simulate a full life in a matter of months. Their correlation with real life is not good, but they are a start point and can be used to compare different approaches.

The points made above relate initially to mechanical systems and components. In the case of electronic systems (Ref. 8.2), the same points are relevant but there are some differences of emphasis compared with mechanical systems. Testing basic functionality, which could be complex and detailed, is the first requirement. In relation to measurement of life, electronic components are less susceptible to wear-out but more susceptible to early-life failures, which are often screened out by a burn-in programme in the case of critical systems. Electronic systems, connectors, and harnesses benefit from development in relation to the installation environment (heating and cooling), vibration, dirt and moisture ingress, and electromagnetic susceptibility (as well as electromagnetic output from the system itself.)

The methods already described apply where the product is developed principally in the organisation's own environment and not released to customers, other than in very controlled circumstances, until it has achieved a high level of integrity.

8.8 Prototypes Not Possible?

A further consideration is what to do if prototype development is difficult or impossible. For example, fully representative prototypes of major chemical plants, bridges, space vehicles, and nuclear submarines are just not possible. Conversely, prototypes of road vehicles and aircraft can be created readily and tested in their normal working environment, or in specialised test facilities where necessary. However, there are practical limitations imposed by costs – commercial aircraft cost tens of millions of dollars.

Other products, such as rail vehicles, find extended running of prototypes more difficult. There are some dedicated rail test tracks, two in Europe, for example, but these have some limitations in terms of length and features. As commented in Chapter 3, when discussing the advanced passenger train, testing new products of this type in normal service with fare-paying passengers is not a good idea.

In principle, several approaches can be taken, driven by the volume of the product and the characteristics of the market:

- Where the product is going to a high-volume consumer market, the expectation is that a reliable product will be available in quantity from the day of the launch. In this situation, the product needs to be fully qualified before the commencement of series production and a pipeline of finished products must be in place before launch.

- In the case of a low- or medium-volume, high-cost product, production can be started slowly, perhaps as a pre-production batch, and products released only to known customers, monitoring performance in the field very closely. Commitment to series production is then made when early-stage problems have been ironed out
- In the case of one-off, or very small quantities, the product will go through a commissioning period when its basic functioning will be achieved followed by a period of 'reliability growth' as problems are overcome. In this situation, customer satisfaction comes from achieving a fully functioning and reliable product in whatever is considered a reasonable time period.

The approach taken is a function of the product's volumes and cost, as well as the established practice in the industry concerned. It is a particularly critical period in that premature release of an underdeveloped product is a guarantee of customer dissatisfaction from which recovery is difficult.

8.9   Physical Test and Laboratory Support Facilities

Physical testing obviously requires an organisation either to have, or to have access to, appropriate test facilities and the means of manufacturing prototype test parts. Their scale is very dependent on the company concerned. Large, well-established organisations will have facilities on a scale capable of dealing with anything from small components to full-scale products such as road vehicles, aircraft, or trains. Start-up companies may rely on university facilities or public laboratories, although they may be able to undertake their own small-scale rig testing.

Whatever methods or facilities are used, test work needs to be backed with a range of critical support facilities. Calibrated instrumentation and data recording systems are an essential part of physical testing. Similarly, data analysis systems play a critical role in extracting and understanding the results of tests. Metrology equipment is needed to measure components before and after test. Laboratory facilities for materials analysis are needed, which can examine any failures of the type listed earlier, literally under the microscope or using other methods, to understand failure mechanisms.

Established practice is for such facilities to carry approval and accreditation beyond the requirements of ISO9001. The specialised standard for laboratories and test houses is ISO/IEC 17025:2005 (Ref. 8.3) which 'specifies the general requirements for the competence to carry out tests and/or calibrations, including sampling. It covers testing and calibration performed using standard methods, non-standard methods, and laboratory-developed methods'. This standard is intended to ensure that a laboratory is technically competent.

8.10 Correlation of Modelling and Testing

The methods described previously, whether in the form of calculation, modelling or testing, depend for their usefulness on their accuracy in representing real life operation. There is no easy solution to establishing this correlation other than building it up over time and thus accumulating valuable experience. The validation methods used will always be an approximation to real-life but they will, nonetheless, still give a new product a thorough examination which will identify almost every potential defect. The opportunity also exists in every development programme to refine the correlation; for example:

- Early calculations can be compared with later more thorough and accurate modelling
- Models can be compared then with physical test results
- Instrumented products can be operated in the field to gather real-life data
- Warranty and customer failure data can be fed back to understand why in-service failures occurred and why they were not prevented

Consciously making correlations of this type will improve the validation process and is an important part of company learning. Connected devices (Internet of Things) will greatly expand the volume of information about the use of products in service – dealing with the sheer volume of data may be the biggest challenge.

8.11 Assessment of Serviceability

Most products have specified requirements for serviceability and repair: at one extreme, some products may be designed (rightly or wrongly) to be thrown away if failure occurs; at the other extreme, complex equipment will require regular servicing and repair throughout its life. The ability to carry out efficient servicing and repair will be built into the documented requirements for these latter products, including, for example, time requirements for specified operations. Confirmation that these objectives can be met should be built into development programmes, as with any other specified requirement.

Quite thorough reviews are now feasible at the design and modelling phase. Three-dimensional CAD models give the ability to explore access to components and the ease of removal or replacement. Virtual reality (VR) modelling adds another level of realism, providing the ability to simulate directly maintenance and repair activities. Such assessments can then be repeated once prototype hardware is available, although this activity should be no more than a confirmation exercise if the simulation work has been thorough. This type of work is best carried out by service personnel, and the findings should be recorded as corrective actions as with any other test result.

8.12   Software Development and Validation

Most engineering products include software, which typically perform control functions and are therefore critical to the operation of the product or system. As an integral part of a system, software contributes as much as any other element of the system to its overall reliability. It does, however, perform or fail in a different way to traditional components, where the emphasis when ensuring reliability is on component and material failure. Unlike these traditional components, software does not fail in a physical way. Rather, it produces unintended outputs when certain combinations of inputs and system states apply. As many of these conditions as possible should be identified and corrected before a product reaches an end customer.

The science of this subject is now very well researched, and whole university departments devote their work to software integrity and, increasingly, security. Their work covers commercial software, such as banking and airline systems, as well as the engineering control systems of interest here. However, in this short section, only a superficial overview of the topic can be provided.

As with other forms of engineering, reliable software derives from a high-integrity design and development process, covering specification of requirements, initial design, software coding, checking on a modular basis as design proceeds, and independent validation before 'full-scale' testing on complete products. These steps are intended to prevent errors rather than having to detect them later.

The number of potential failure paths in software is a function of its complexity. Detecting failures will therefore be more effective the longer that software is run and the wider the range of conditions for those runs. When errors do occur in software, they are not always easy to diagnose, and they can be difficult to distinguish from hardware problems. However, software errors should repeat – there is not the same variability in coding as there is in manufactured components, and software does not degrade or wear out like physical components.

In exceptionally high-integrity control systems, it is accepted that absolutely fault-free software is unattainable and therefore some level of redundancy may be built in. One path in a multiply redundant control system may then incorporate software written by an entirely independent team to the other paths in the system. This way, the error-generating combination of inputs and system state will be handled differently in the back-up system.

Software validation therefore needs to be written into a product's overall validation plan, including physical testing in as wide a range of conditions and for as long a period as can be afforded.

8.13   Reliability Testing

Reliability, as a defined term, relates to the expectation that a product or system will operate as intended when called upon to do so. The reliability of existing products operating in the field is often measured objectively using field service data. Following on from this, numerical reliability targets are often set for new products expressed, for example, in terms of failures per year, or reliability percentage per mission, or mean time between failures (MTBF) – see Figure 8.4.

A point to bear in mind is that there are different understandings of what constitutes 'failure'. To some, a fastener coming loose and requiring tightening is a failure. To others, the definition relates to the overall failure of a system to function. Different industries have different practices in this respect.

Specifying reliability, whatever measure is used, does, however, raise a fundamental question. Deliberately designing a product to meet a specific numerical target is just not possible; reliability performance is more of an outcome of the development process than a property of the product. It is not like such

Abbreviation	Title	Description
	Probability of failure	Likelihood, e.g. in percentage terms, of failure over a given period or for a particular mission, e.g. 0.1%
	Reliability	(1 – Probability of failure), or probability of performing a function over a given period or during a particular mission, e.g. 99.9%
MTBF	Mean time between failures	Mean time between successive failures on a repairable product or system, e.g. 500 hours
FPHV	Failures per hundred vehicles	Number of failures or problems experienced over 12 months on 100 vehicles
	Availability	Proportion of the occasions when a product or system is capable of functioning when called upon to do so, e.g. 99.9%
PFD	Probability of failure on demand	For one-shot systems, e.g. an emergency shutdown system, the probability that the system will fail to operate when called upon to do so

Figure 8.4 Common measures of reliability.

properties as mass or top speed. It could be argued that, if reliability could be specified and designed for, then the modes of failure must be known and could therefore be avoided in the first place, which is clearly not the case.

The development and validation methods described previously have the basic aim of identifying defects in the design so they can be corrected before release to the customer. The more defects that are found and corrected, the greater will be the reliability of the product when operating in the field. This statement presupposes that defects are inherent in the design, which is partly but not entirely the case. In addition, the manufacturing process is equally capable of generating defects although a good design will simplify manufacturing and thus reduce the chance of error.

The question then arises as to whether reliability can be objectively measured as part of the development programme. Clearly, this would have to be done in the latter stages of any programme when the bulk of issues had been addressed. It would also ideally use production, rather than prototype, material with all the normal variability that production processes entail.

Within these constraints, a reliability measurement programme is possible in principle. If the intended failure rate of the product is of the order of one to two defects per year (see earlier comment in Chapter 7), then generating meaningful data would require a minimum of six to eight products built to the same standard and operating for 6 months plus, unless some way of accelerating usage can be found. With relatively inexpensive items, such as domestic appliances, 20 to 30 items could feasibly be tested, giving better results from a statistical viewpoint.

A further approach would be to undertake reliability trials on individual components or subsystems. They can be run in significant quantities in test laboratory. It will not be completely representative of field operation, but useful information will be generated and results such as failure mechanisms will be generated quickly.

Fortunately, there are methods of deriving reliability estimates from very small sample sizes. A widely used method (Ref. 8.4) was developed originally by the Swedish engineer Waloddi Weibull, who first presented his ideas in 1939 and his definitive paper in 1951 to the American Society of Mechanical Engineers, including seven examples of where it could be used. His approach met with some controversy but has proved valuable in the field of engineering, as much as anything for its relative simplicity and clear presentation – it is based on a log-log plot of cumulative failures against time or number of cycles. Figure 8.5 shows an example.

Technically, Weibull's method relates to one failure mode at time, rather than the mixed failure modes that would be experienced on a complex product. There are related methods (developed by J. T. Duane and Dr Larry H. Crow), again using log-log plots, which are more accurate when considering multiple failure modes. These methods give the development engineer the ability to gain

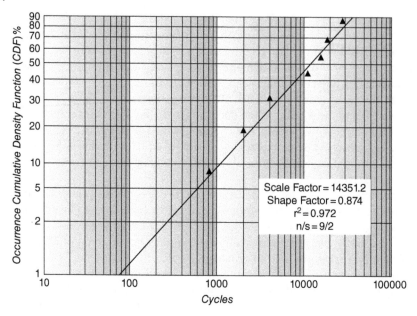

Figure 8.5 Typical Weibull plot.

some early indication of numerical reliability and hence to judge how a product will be received in the field.

In conclusion, reliability testing does have a place in terms of understanding the maturity of a product and whether it will have adequate reliability for the marketplace. It could continue to be measured, either from data from in-service products or by testing of production batches, as a form of quality oversight. It will, however, measure the outcome of the development process rather than being the means of achieving the target reliability.

8.14   Corrective Action Management

The purpose of all the development and validation work described above, whatever form it takes, is to generate knowledge and learning about the performance and life of the product. This is only of use if something is done with the information, which, in turn, implies a conscious process of managing the learning that has occurred. This point was made earlier in Chapter 2, where some of the principles of managing learning were made. In summary, the learning or corrective action process should record the following:

- A description of the problem encountered or the potential improvement that could be made
- A record or analysis of the problem and what detailed information or data are available
- An assessment of the root cause of the problem, based on the data about it
- A note of the potential solution (when first recording the problem)
- A note of the corrective action planned
- A record of the problem having actually been closed out in subsequent phases of work

This does not need to be a complicated or bureaucratic process; the main point is ensuring that all opportunities for learning are acted upon and not lost or forgotten. (Early signs of most problems arising in service can be found in earlier stages of development. And service records are another source of learning for future projects.)

8.15 Financial Validation

At this point of development, major financial commitments will either have been made or will be close to being made. Although not part of the technical validation of a new product, a detailed financial review of the project should be undertaken at this stage. This needs to be based on obvious parameters such as sales volumes, sales revenue, and product costs but also needs to include one-off costs such as tooling, investment, development costs, launch costs, and (often forgotten) working capital and spares inventories.

The most rigorous form of analysis at this point is in the form of a discounted cash flow, using the organisation's weighted cost of capital as the discount factor. Figure 8.6 shows an example of such an analysis with cumulative cash flow (units could be $m's or £m's) plotted against time, in years.

This particular example has a period of 2 years prior to launch and then 7 years afterwards and uses an 8% discount factor, which might be considered low. Every company will have its own criteria for judging what is considered to be an adequate return and, of course, different projects could be compared using this methodology.

The key point is the compilation of a time-based inventory of all future costs and revenues. Although the example and the text are based around a volume product, the same principles can be used with one-off or very low volume solutions. It can also be used where ownership of the product is retained by the manufacturer and revenue arises from lease or from sale of the 'effect' of the product. In these instances, of course, cash flow is delayed, relative

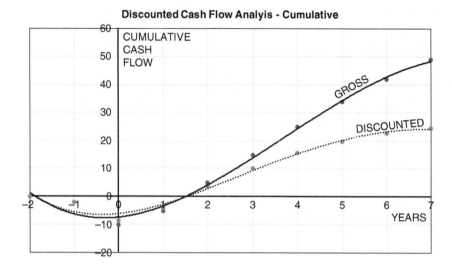

Figure 8.6 Example of discounted cash flow analysis.

to a straight sale, so a time-based analysis is even more important to ensure financial viability.

8.16   Concluding Points

This chapter has described the processes for ensuring that technology and product development effort delivers a robust and reliable end result. This objective is partly achieved by selecting the right concepts in the first place but is more dependent on the detailed execution of thorough development programmes. These programmes would ideally be based around an organisation's existing development and sign-off codes, which represent that the organisation's accumulated learning and experience in this field. Programmes would also be driven by risk analysis, identifying areas of particular concern, especially where new technology, new markets, or new operating regimes are coming into play. The latter three are the areas that could be problematic as they are, by definition, outside previous experience. For start-up companies, all experience is new and, hence, developing a new technology or product to the point where it is reliable is even more challenging.

Development and validation activities are undertaken by a combination of engineering calculation work, modelling and simulation, and physical testing, in approximately that order. The more thorough and the earlier the calculation and modelling work, the less physical testing, which is expensive, will be needed. This is better also for the project in total, as other planning work will benefit from having better-researched early stages. However, there is no substitute for some level of physical testing. In a complex product, physical prototypes are the opportunity to see all the components of the product working together – something that is difficult to simulate. Validation and sign-off, including regulatory approval, are normally based on physical products. Physical testing does need access to competent facilities that require extensive support facilities for instrumentation, data gathering and analysis, and failure analysis.

The activity of development is very much one of learning: understanding how a product performs and how it fails, which, in turn, will provide an understanding of what margins exist in normal operation. The work therefore firstly confirms that the product or system meets the performance expected of it in terms of its functionality, including any regulatory requirements that are relevant. This information should be understood for a range of operating regimes in terms of duty cycle and environment, for example, and not just for everyday usage.

The work secondly looks at the life of the product for which various methods of accelerated durability testing exist. The methods of compressing a number of years of life into a short period are not perfect, but they will flush out most problems.

It goes without saying that development and validation programmes need to be carefully planned, with the activities following each other in the right sequence. The learning points generated by the work need to be recorded and follow-up confirmed.

In the later stages of development programmes, numerical measurements of reliability can be made if a statistically significant (this need not be high) number of products can be made by production methods and operated in realistic conditions. It should be noted that a product's reliability is set by the thoroughness of the development programme and is not an inherent 'property' of the design.

Financial validation should also be undertaken at this point and methods for doing so are described.

The points above relate mainly to products that are produced in some quantity and where there is a clear distinction between prototypes and production items. Some engineering products are produced either as one-offs, e.g. process plants or bridges, or in very low and expensive quantities, e.g. submarines or space-craft. In these situations, an even greater emphasis is placed on calculation and modelling work. The physical product is then the final item which must go through a period of commissioning and shake-down. Customer satisfaction in this situation comes from engineering solutions which perform well from the outset and where the commissioning process is effective in finalising the solution.

References

There are relatively few publications relevant to this chapter.
Neville Shute's autobiography includes some interesting material from 80 or 90 years ago, including the forerunner of finite element methods, performed manually:

8.1 Shute, N. (1954). *Slide Rule, the Autobiography of an Engineer*. Norway: Heinemann.

This book provides a substantial overview of reliability engineering as applied to electronic systems:

8.2 Swingler, J. (2015). *Reliability Characterisation of Electrical and Electronic Systems*. Woodhead Publishing.

Engineering test laboratories should be set up to operate within this international standard:

8.3 ISO/IEC 17025:2005 *General Requirements for the Competence of Testing and Calibration Laboratories*

Weibull statistical methods are useful where there is a relatively small number of test samples:

8.4 McCool, J.I. (2012). *Using the Weibull Distribution: Reliability, Modelling and Inference*. Hoboken, NJ: Wiley.

9

Engineering Delivery

9.1   Introduction

This chapter is concerned with the final stages of technology and product development programmes. This is where the ideas that emerged several phases previously make their way into production and operation.

Taking a somewhat narrow view, and perhaps understating all the creative thinking that has gone previously, the final output of these development programmes is information. For the parent organisation, this is the information from which reliable new products can be manufactured, sold, and operated. This chapter describes how this information is finalised and delivered to its users.

As has been stressed elsewhere, this development and provision of information is not an 'over-the-wall' affair but a collaborative exercise where the product development engineers work closely with people in:

- Manufacturing
- Purchasing and supply chain management
- Service and product support
- Marketing

The engineers are the custodians of the product information, developed jointly by this group. It represents a critical, intangible asset of an organisation, despite the fact that it will not be financially recognised on that organisation's balance sheet. As an aside, investments in intangible assets now outnumber conventional asset investments in at least three countries (Sweden, United States, and United Kingdom) – see Ref. 9.1. They also have the advantage of being far more scalable – i.e. they can be used at increasing volumes, far more easily than physical assets.

This phase of activity corresponds to technology readiness levels 7 – 9, as described in Chapter 3, and manufacturing readiness levels from approximately MRL 5 and upwards. In terms of the stage-gate model described in Chapter 5, the phase begins after the third review point and concludes at the fifth.

Managing Technology and Product Development Programmes: A Framework for Success,
First Edition. Peter Flinn.
© 2019 John Wiley & Sons Ltd. Published 2019 by John Wiley & Sons Ltd.

This phase consumes the majority of the cost of a development programme. It is where the bulk of the detailed designing, modelling, prototype manufacture, and test work are carried out. Earlier phases will involve all these activities as well, but at a much less intense level. At the conclusion of the phase, the various types of information will be in their final form, based on the feedback and learning from, for example, test work, manufacturing reviews, supplier discussions, and field trials, where these are relevant.

Given the number of people involved, perhaps across multiple organisations and locations, the number of engineering components, and the amount of data, it is a phase of work that does need significant structure and planning. Everyone needs to be clear about what is required and by when, in terms of major deadlines. However, there also needs to be a degree of flexibility, as learning and iteration are still prevalent to a degree; earlier, pre-TRL 7 work should have given a solid foundation. There should not be any show-stopping technology problems if the earlier work has been done thoroughly. However, detailed changes and vital improvements will come out on a daily basis.

The organisation undertaking this work is sometimes referred to as the company's design factory. There is an element of truth to this description, but factory management techniques, with one or two exceptions, definitely do not work in an environment where learning and iteration are still very much to the fore.

9.2   Forms of Information Output

The output from this phase of work fully defines the product, enabling it to be built, operated, serviced, and retired. The engineering information includes:

- Digital geometric models, usually in three-dimensional form and from which conventional drawings can be derived.
- Bills of material, listing the parts, their formal part numbers and issue level, their quantities, and their assembly hierarchy. From this listing, manufacturing and supply sourcing routes can be derived and production scheduled. They also form the basis of cost models, weight estimates, and any other properties that require a complete parts-listing.
- Written specifications for material properties, for example, or to define the performance requirements of parts, assemblies, and systems sourced as proprietary items, designed by their supplier.
- Product acceptance test and commissioning instructions.
- Regulatory documents, certificates and conformity documents.
- In some instances, safety cases or similar detailed documents to enable safety approval or formal acceptance by regulatory authorities.
- Physical operating models or prototypes that can be used for demonstration purposes.

- Operating, servicing, and fault-finding instructions.
- Disposal or end-of-life instructions.

This information then forms the basis of manufacturing and procurement plans, which represent a further and substantial area of documentation including: production processes, factory layouts, quality management processes, and supplier management, to name a few, and not covered in any detail in this book. Figure 9.1 shows the scope of engineering information and the 'delivery system' for providing it to the company, suppliers, dealers, and customers. Not all companies will follow precisely the same pattern as shown on the diagram; some, for example, might not involve distributors. However, the principles still hold.

It is important to note that there should be one master source for each of the pieces of information defined above, probably with one controlling individual and all with formal version control. There are still, unfortunately, companies that run multiple bills of material, for example, which is a recipe for confusion and wasted effort. When electronic, international access to master databases is readily arranged, the multiple and unofficial lists, common some years ago, have no place in modern engineering. The basis of these disciplines should be created from the early stages of technology development – it is not just the province of the later, delivery stages of product development projects.

9.3   Connected Products – Internet of Things

Information connectivity to and from the manufacturer to operating products is becoming an increasingly important aspect of engineering, adding a further dimension to the diagram above. As well as designing sensors and data transmission systems into products, it will also require the portfolio of product information produced by the manufacturer to include software for data analysis. This, in turn, then prompts the need for guidance and instructions about how to interpret the data, what constitutes a faulty condition, for example, and what supporting service the customer then needs as a consequence. Internet of Things (IoT) opens up all sorts of interesting possibilities for which there must be a clear business case but which has the potential to add a new and valuable dimension to a manufacturer's offering.

9.4   Detailed Design

Concept development will have generated predominantly schematic layouts of the proposed technology or product. There will probably also be some individual component details, especially where tests on real items have taken place. The detailed design phase takes this starting point and produces complete engineering information, such as drawings, on all individual components, as well as how they fit into subsystems and systems. Whilst the concept

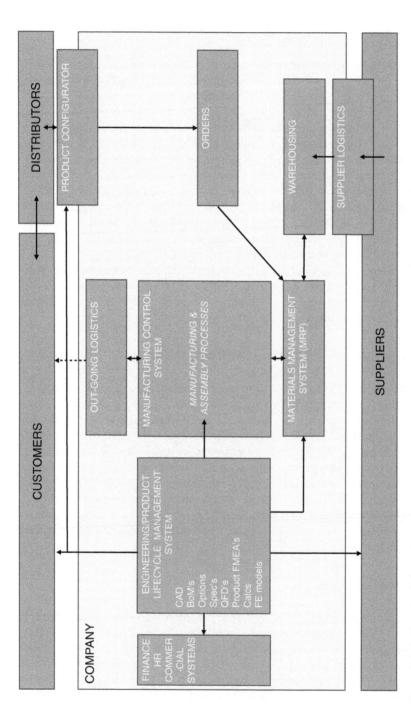

Figure 9.1 Engineering information.

fixes the cost envelope for a new product, the detailed design fixes costs and manufacturability at an item-by-item level.

In terms of cost, the difference between good and bad design can easily be $\pm 20\%/30\%$. Simpler designs provide cheaper and more elegant solutions, as well as greater reliability. Most component designs are attempting to balance a range of factors – function, cost, weight, appearance, and manufacturability are some examples. Basic calculations such as stress analyses will be used to evaluate components at this stage.

Components also interact with adjacent parts to which they are attached or are connected or which they must avoid contacting. This obviously places limitations on how the design can be achieved, and there is a constant juggling act to synchronise component designs. Once complete, designs can be assessed in terms of complexity indices, which look at the number of parts versus the number of functions in an assembly – see Stuart Pugh *Total Design* (Ref. 9.2).

The work relies, in particular, on good, basic engineering practice to generate workable, efficient solutions that meet the needs of customers and other interested parties.

9.5 Handling the Interfaces

Based on the detailed design work above, this phase of work is particularly characterised by the generation of high volumes of new and detailed information. This is partly a function of the greater number of people involved, and hence the quantity of work completed and partly the sheer breadth of activities being undertaken. New information and new issues will be continually generated by detailed design work, analysis and modelling, testing, manufacturing development, and supplier work. This is an inevitable and, for some, inconvenient truth. There would be no point in doing the work if this were not the case. How well an organisation deals with this situation is the acid test of its product development capability.

The focus is then on the decisions about how to respond to this new information. Some data may simply confirm what had already been assumed or planned, in which case no follow-up action is required. Other information may suggest changes to avoid problems or to improve the product or its manufacturability, This, in turn, leads to discussions and decisions about:

- Whether and how to change the product definition
- Whether to re-run the work, such as tests, which identified the potential change, with modified material
- What to do with material that has already been ordered or made to the previous standard
- What impact the potential change might have on costs, investments, and timescales

Each decision should be taken on its own merits, and there are no prescribed rules as to how this should be done, except to ensure that a full picture is assembled of the relevant facts before deciding what to do and that the decision should be taken collaboratively.

9.6   Cost of Delayed Programmes

A critical point to bear in mind in these situations is the scale of the potential cost of delay. In Chapter 4, attention was drawn to the high profitability and hence high financial returns from technology and product development programmes. The obvious consequence of a delayed product launch therefore is a high financial opportunity cost. Some companies can quantify this cost quite easily, whereas for others it may be less straightforward. Some examples are:

- *Value of output well known.* An example might be an oil refinery. If a refinery is being constructed with capacity of, say, 500 000 barrels per day, a quick calculation at $50/barrel suggests $12.5 m per day material going in and margins presumably measured in $100 k's or $m's per day. In this situation, delay is very expensive.
- *Penalty clauses apply.* Some contracts have liquidated damages clauses with penalties for each week of delay (contracts in this form are a sure recipe for poor quality and strained relations).
- *Seasonal product.* Some products must hit summer or Christmas markets. Even a few weeks' delay can result in lost sales and profitability through loss of a season's sales.
- *Loss of profitability.* New products are invariably planned as sources of new profitability. The value of delays can be quantified in terms of the lost profit opportunity as well as the additional cash flow exposure as a result of a longer development programme.

Every company could estimate the cost of project delay, if only in approximate terms. Such estimates invariably point to a preference for maintaining timescales, provided the product meets an acceptable sign-off standard (see below).

9.7   Planning and Decision-Making

The approach that can be taken to planning and managing all phases of technology and product development programmes has already been described in some detail in Chapter 5. The delivery phase of such programmes, the subject of this chapter, is the most complex, organisationally, and involves the most resources, much of which could be spread amongst different organisations and locations.

It therefore requires careful planning at the appropriate level of detail. Specifically, everyone involved needs to know what they have to do and by when. This does not mean, however, that planning must be overly complicated or bureaucratic. In fact, simplicity of planning is an essential aid to effective and clear communication to a large team.

The starting point is a high-level master plan showing the overall timescales. This is likely to be driven by:

- Timing of the release of early engineering information, sufficient to start manufacturing planning and procurement
- Procurement timescales for long lead items
- Facility and investment lead times
- Lead times for establishing sales and service networks
- Lead times for establishing data networks
- Final engineering release based on validation work
- Judgement on the extent to which the procurement and investment activities can be overlapped with the engineering development

Each element of the project then needs its own subplan with key milestones. Those for engineering, manufacturing, and procurement need to go down to a part-by-part level (so a complete bill of material is vital). In most cases, the plan needs simply to be a list that can be ticked off as progress is made, rather than a complex network.

The programmes for other parts of a large project can take the form of milestone plans, linking back to the master plan.

Day-to-day management can then be handled at a team level where there will be constant review of progress, issues, learning points, and priorities. This topic has already been covered in Chapter 5 in Section 5.13, 'Monitoring Small Projects or Sub-Projects'. Above the level of the individual team, there should be a tight structure of less frequent, but timetabled, formal meetings, perhaps culminating in stage-gate reviews. The aim always is to contain problems and deal with them flexibly and imaginatively.

Don Reinertsen, in his book *The Principles of Product Development Flow* (Ref. 9.3), draws a detailed comparison between engineering and military planning, as specifically practised by the US Marines. In the heat of war, the generals cannot control each individual unit or soldier directly. They lay out the overall plan, especially the objectives, and rely on the units having a good understanding of these objectives to take the initiative as the battle proceeds. In the military situation, the right balance must therefore be struck between central coordination and decentralised responsiveness. (See also Ref. 9.4.)

In engineering, new information, which is constantly emerging, is first visible to the front-line engineers, who should adapt their tactics to respond to the new situation. The engineers should be trained and developed to work in this way. They must have the trust of the organisation to respond in the right way, plus the knowledge that the backup is available should problems escalate. Engineering therefore follows the same principles as the military, but with only reputational injury rather than physical injury at stake!

9.8 Specialised Resources

All engineering programmes rely to an extent on specialised resources in the form either of people or facilities. More specifically, these resources could include:

- Individuals with special knowledge and skills for development work
- Senior and experienced staff for technical reviews
- CAD facilities and software
- Computing facilities and software for analysis and modelling (computer-aided engineering, CAE, resources)
- Manufacturing capabilities for making and assembling test components
- Test and data analysis facilities
- Senior managers, required to make decisions or formally sign-off proposals

Within small organisations, with only one or two projects, some of these points may be less of an issue, but such organisations will depend more on external suppliers with whom they will have limited leverage. Large organisations will possess internally most of these capabilities but will be handling multiple projects with different priorities, competing for resources with each other.

The result of these bottlenecks is stoppage. It can be thought of as queues of part-finished engineering work-in-progress (WIP). It is not physically visible, as is the case in a manufacturing plant with physical inventory, and it is not recorded as WIP in an organisation's accounts. Hence, it not managed directly – remember the adage, 'What gets measured gets managed'.

If anything is measured in this situation, it is more likely to be utilisation of resources, which actually has the effect of increasing bottlenecks. (Queuing theory shows that, as utilisation increases so do queue lengths, alarmingly so as 80% + utilisation figures are achieved – see Figure 9.2.)

The point is that bottlenecks of specialised resources can have a huge effect on product development throughput times and in slowing down the important feedback loops, which are at the heart of the process. In many situations, it is an invisible and unmeasured constraint on progress.

In terms of what can be done, the following are suggested:

- Understand that this problem exists.
- Try to increase the level of resource in known bottlenecks – easier said than done.
- Reduce discretionary workload on bottlenecks and hence reduce their utilisation.

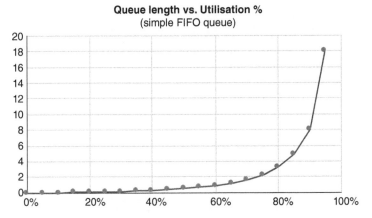

Figure 9.2 Queue length versus percentage resource utilisation.

- Find ways of reducing the work content of tasks assigned to bottlenecks, concentrating on what absolutely must be done by specialist resources.
- Train general engineers to be capable of doing at least some of the bottleneck work.
- Book external facilities well ahead of the point of need.

The reason that projects can often be accelerated when given management attention is that those projects jump the queue in these bottleneck areas. The real test of product development management is whether cycle times for all projects, not just the priority ones, can be improved.

9.9   Flow of Information

The topic of specialised resources highlights one area where a direct parallel can be drawn with lean manufacturing. Engineering work is rather like material in a (poorly organised) factory, criss-crossing from one workstation to another. A plot of the time components spent travelling, waiting, and being worked on will show limited value-added time but much wasted time. The same principles can be applied to engineering WIP. One of the advantages of project-based organisations (see Chapter 5) is the tightening of these flows. Drawing on manufacturing practice, it is perfectly possible to set up classic U-shaped cells for engineering work, as in Figure 9.3.

The membership of such cells can be expanded, e.g. to include analysis or test engineers, or contracted to suit the situation. They could be staffed permanently or people could get together for a few days a week then return to their normal base (this is one way of creating a hybrid between the functional and project style of organisation – see Chapter 5).

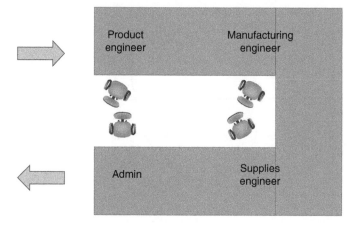

Figure 9.3 Product development U cell.

9.10   The Importance of Good Systems

Whilst face-to-face contact and discussion is the most effective means of taking product development forward, it needs to be underpinned by good systems, especially in two areas:

- *Product data.* Design information, parts lists, specifications, product acceptance tests
- *Engineering process management information.* Progress tracking, change control, and learning points

The key point about the first of these is that the most up-to-date and comprehensive product and process information should be widely available to the development team – something that is entirely possible, technically.

In terms of process management, Chapter 2 (Engineering as a Process) described two processes that are critical to the development process – change management and recording problems or learning points. Both need to be tracked so everyone is clear about progress, what is still open, and what has been closed.

9.11   The Role of Standards and Design Codes

Standards and design codes play an important role in ensuring the integrity and reliability of new designs. In many industries, longstanding international standards exist that in effect represent the collective wisdom of generations of engineers working in the field concerned. Examples can be found in construction, pressure vessels, process industries, automotive, and aerospace, to name but a few. They are particularly important in relation to integrity and factors of safety where long experience has shown what will and won't work safely. They may also underpin certification or regulatory approval.

The derivation of such standards is often therefore empirical, based on observation rather than logical calculation. This reflects the fact that patterns of usage, duty cycles, and abuse are often difficult to define precisely although modelling and calculation is used to back up more intuitive conclusions. There are also more general standards on test methods, materials, laboratory methods, terminology, and management systems. Organisations such as ISO, SAE, ASME, BSI, and CEN (see Glossary) have the most comprehensive libraries of standards.

Where standards already exist, and there is no field of engineering where this is not the case, companies developing new technologies or products should clearly use them. However, when it comes to the detail of new configurations of technologies and products, new methods of design will be generated by the originating organisation. These new product-specific methods represent critical intellectual property that should be written into company design codes ('how-to' documents) that enable the technology to be reproduced. Such codes will be a mixture of calculation methods, assumptions, test methods, and relevant existing codes. Taking a simple example, a design code for a cooling system might include:

- Operating environment assumptions, e.g. temperature and humidity range
- Calculation methods or models, which could be coded as software
- Materials standards
- Tolerance to off-design conditions, e.g. heat exchangers partly blocked, coolant level low, sensor inaccuracy, incorrect antifreeze, or inhibitor concentration
- Safety shut-down approach
- Test and sign-off methods
- Relevant international standards
- Maintenance assumptions
- Effect of loss of electrical power during normal operation

Well-established companies have libraries of design codes and standards of this type representing an often-underestimated repository of company learning. New companies should think in terms of developing new design codes to document their learning, something investors also should encourage.

9.12   Tracking Product Cost and Investment

The points above are mainly concerned with the technical development of products and their performance. But as stressed in Chapter 4, new technology

	Current Estimate	Previous Estimate	Target
Material			
Direct labour			
Total direct costs			
Product-specific investment			
Commentary			

Figure 9.4 Tracking of product cost information.

needs to be commercially as well as technically successful if it is to be of any value. An important part of the programme, therefore, is to ensure that all financial targets (product cost, labour content, investment, and operating costs) and sales volumes are met.

This is best done by tracking on a part-by-part basis from the earliest possible stage – see Figure 9.4. Apart from ensuring accuracy and completeness, this also devolves responsibility, enabling costs and related topics to be managed alongside 'engineering' parameters.

Continuous reporting of financial data and projected sales volumes enables problems to be highlighted early so that management can take action.

9.13   Knowing When to Stop

There then comes a point when improvement may have to stop. If the development programme throws up issues that simply must be dealt with, e.g. to achieve legal compliance or to avoid problems in service, then delay or additional cost may have to be accepted. Where there is some discretion about potential changes, then the high cost of delay comes into play, and it is often better to continue with what already exists rather than incorporate further improvements.

Seasoned engineers frequently use the expression 'the better is the enemy of the good' for guidance at this point. They probably don't realise that, in doing so, they are drawing on the wisdom of the past. In Shakespeare's *King Lear*, the Duke of Albany said, 'Striving to better, oft we mar what's well' and Voltaire ascribed his statement that '*le mieux est l'ennemi du bien*' (i.e. 'the best is the enemy of the good') to an Italian sage.

One of the important roles of the project leader in this situation is to bring product development to a halt and move into the production phase. In some companies, there may be a formal process for doing this, as described next.

9.14   Signing Off the Product

A good discipline in the concluding stages of a product development project is to have a formal process of sign-off and 'release'. The objective is to confirm that the product, having completed its programme of development, is in a fit state to be manufactured and sold. The process revolves around doublechecking that all requirements have been met, all concerns have been answered, and, as far as possible, nothing has been missed so there are no loose ends.

This could mean looking through all the risks identified, analysis reports, test reports, design-for-manufacture studies, and corrective action records to confirm that all have been cleared. To a certain extent, systems can be developed for keeping a running record of these points, and this is perfectly possible when validation work has shown a clear pass or fail. However, there will be situations when requirements have not been fully met and development work or design-for-manufacture has shown a trade-off between requirements, rather than a black-or-white result. Alternatively, there may have been solutions proposed to previous problems, which are still undergoing durability tests. Customer requirements also develop over time: new ones may emerge and some previous ones may reduce in importance.

Judgement may therefore need to be exercised as to whether to continue to enter the production phase or whether to pause. The ideal decision would be 'OK for production and sale' but decisions could be in the form of 'OK to start production but not released for sale' or 'OK to produce a certain quantity'. The worst outcome is 'Delay production for xx weeks'. These are topics that could be taken by a stage-gate review board, based on the input from the project's engineering leader.

9.15   Examples of Good and Bad Practice

To put the foregoing into context, Figure 9.5 summarises the key points about two contrasting projects from the author's direct experience. They are both real-life projects involving relatively complex, engineered products. As it happens, both were projects won through a bidding process to single customers, rather than direct sale to multiple customers, but this is not relevant either to the way the projects were run or to their outcome.

The contrast between these two projects, executed within a few years of each other, could not be clearer, in the eyes of both the suppliers and the customers. The first 10 points above are the inputs to the process and 11 and 12 are the outputs. This suggests that careful attention to these 10 points is the key to developing a strong capability in the delivery cycle of new product development. Ref. 9.5 also provides some useful guidance.

	PROJECT A	PROJECT B
1. Company drivers	Driven by quality and processes	Driven by deadlines and costs
2. Company culture	Co-operative, shared goals	Inter-departmental rivalry, over-the-wall mentality
3. Market need	Very clear need identified and approved by customer	Very clear need identified and approved by customer
4. New technology	Limited new technology; new elements pre-tested	Much technology new to industry but used in other industries; very limited pre-testing or early manufacturing work
5. Product definition	High-level performance specification from client	High-level performance and detailed design, performance and other specifications from client, much information, some contradictory
6. Design information	Single design authority, common bills of material, common technical specification, built to drawing. Cooperative change management process	Single design authority but, multiple bills of material, technical specification mainly from client, design continually questioned by client, major manufacturing deviations from drawing
7. Calculation, modelling & simulation work	Thorough engineering analysis	Thorough engineering analysis
8. Physical testing	Comprehensive programme with multiple prototypes resulting in many detailed changes; followed by competitive trials by client	Very limited programme
9. Manufacturing planning	Detailed design-for-manufacture reviews	Very limited reviews, poor adherence to drawing
10. Supplier management	Cooperative and constructive relationship	Contractual relationship
11. Service performance	Minor problems in service	Multiple service problems; major, lengthy and expensive programme of rework
12. Customer satisfaction	High satisfaction, product profitable	Very low satisfaction, disputes about responsibility, legal claims and counter-claims, compensation claims. Unhelpful newspaper publicity. ***But product eventually performed well!***

Figure 9.5 Outcome of two engineering projects.

9.16   Concluding Points

This chapter has covered the detailed engineering phase of new product development projects, including the preparatory stages for production, which are important to developing the manufacturability of the product. At the end of this phase, a commitment is made to series production. It is the most resource-intensive part of new product programmes, involving all areas of an organisation who must cooperate in bringing the product to successful fruition.

The output of the engineering programme is technical information – drawings, parts lists, specifications, and regulatory material, for example. Describing this output merely as information underplays its significance, given all the effort and creativity that has gone into it. However, this information is a major corporate asset, easily comparable with physical assets in terms of its importance and therefore deserving adequate protection.

The success of this phase of work is very dependent on what has gone previously. New technology should have been matured to at least TRL 6. Similarly, there should be a strong and well-researched marketing requirement and a good business case. Without this foundation, problems will certainly arise in terms of cost, quality, and timescales.

It is a highly cooperative activity. Given the number of people involved, planning and coordination are critical, but both need to be done in a way that recognises devolvement of decision-making (it is impracticable to centralise this, and it would slow the cycle substantially). Product development work throws up problems, issues, and learning points at an alarming rate, but the bulk of these can be dealt with at the level of the team through frequent interactions and cooperative working. Physical proximity makes this much easier. Some problems, though, may have to be escalated, so structures to facilitate this should be in place.

A particular focus should be placed on 'bottleneck' activities, which usually revolve around specialist, or difficult to access, people and facilities. They can dictate the overall pace of progress, involving long but invisible queues of work waiting to be done.

Product cost, investment, and sales volumes should be tracked and updated throughout the programme to make sure that these important parameters are where they should be.

At the end of the phase, the wish to improve the product still further may have to be curtailed but, at the same time, it is sensible to conduct a full, formal review to confirm that the product is fit-for-purpose.

Finally, a brief summary is given of two contrasting projects from the author's direct experience. The 10 critical features of these two projects form a good guide to what makes a project successful or unsuccessful.

References

This reference describes the growing importance of intellectual property in modern economies:

9.1 Haskel, J. and Westlake, S. (2017). *Capitalism Without Capital: The Rise of the Intangible Economy*. Princeton, NJ: Princeton University Press.

This reference describes how detailed design can be achieved:

9.2 Pugh, S. (1991). *Total Design: Integrated Methods for Successful Product Engineering*. Reading, MA: Addison-Wesley.

Don Reinertsen, amongst other points, draws a comparison between military work and engineering programmes.

9.3 Reinertsen, D.G. (2009). *The Principles of Product Development Flow*. Redondo Beach, CA: Celeritas Publishing.

His first book, originally published in the 1980s, is one of the first books to look at the product development process in the round:

9.4 Smith, P.G. and Reinertsen, D.G. (1997). *Developing Products in Half the Time Second Edition New Rules, New Tools*. New York, Wiley.

Westrick and Cooper's book has a plethora of practical guidance, based on experience, about how to run engineering programmes:

9.5 Westrick, R. and Cooper, C. (2012). *Winning by Design: Practical Application of Lean Principles for Transforming the Speed to Market, the Quality, and the Costs of New Product Development. Create Space Independent Publishing Platform*.

10

Funding the Programme

10.1 Introduction

At risk of stating the obvious, money is required in substantial quantity to fund all stages of technology and product development activities. And the rate of spend increases as the technology advances in maturity along the technology readiness level (TRL) scale, usually by several orders of magnitude. In the early stages of a project, there might be just a handful of people involved in the development work, but even this will cost tens of thousands of pounds or dollars per year, if properly and fully costed. In the later stages, the potential rate of spend could be up to a million per year, even for a very modest programme and, of course, tens or hundreds of millions a year for a large undertaking, such as a new automotive or aerospace programme.

There are multiple potential sources of funding for company growth. Each source has its own characteristics, and there is overlap in terms of what each category might cover, depending on the approach taken by individual funding organisations. Most engineering companies use a combination of those listed in Figure 10.1.

Internal sources are always the starting point, including an individual's own time and money in the case of a start-up through to cash flow generated from sales with established companies. In the latter case, companies have to generate enough cash to at least maintain the competitiveness of their current products.

The *friends and family* category is self-explanatory. It is the next port of call for new organisations at the very early stages of their development, with its obvious limitations and pitfalls, but an essential ingredient in the development of very new organisations. Other sources will not lend money unless there is some personal money at stake.

Once beyond this stage, there are several private, professional equity funding sources, ranging from angel investors and crowd-funding through to private equity companies. Their approaches vary widely in terms of the levels of funding that they can offer and the point in the company development cycle where

Managing Technology and Product Development Programmes: A Framework for Success,
First Edition. Peter Flinn.
© 2019 John Wiley & Sons Ltd. Published 2019 by John Wiley & Sons Ltd.

		Early-stage company or start-up	Growing company	Established company
INTERNAL	Own resources	✓		
	Own cash flow		✓	✓
EXTERNAL – PRIVATE	Friends and family	✓		
	Angel investors	✓	✓	
	Venture capital		✓	✓
	Private equity			✓
EXTERNAL – PUBLIC	Crowd funding	✓	✓	
	Bank lending		✓	✓
	Initial public offering (IPO)			✓
	Government and other public sources	✓	✓	✓

Figure 10.1 Sources of funding for development work.

they are comfortable to participate. In all cases, they will scrutinise very closely the risk/reward profile and the business plans of candidate organisations.

This then leads to wider equity funding, which could come through some form of initial public offering that would bring into play a range of institutional investors. These sources of funding are mainly relevant to organisations that are quite well-established in terms of products, markets, and sales revenue but are looking to expand.

Finally, multiple forms of grants available from regional, national, and supra-national (EU) public programmes that specifically target technology development, this being seen as a means of encouraging economic development. Allied to these, a number of countries have national networks of engineering facilities, established by public funding, to support the practical development of new technologies.

Each of these sources is discussed in more detail in the remainder of the chapter, with emphasis on early-stage funding, which will be of most interest to the entrepreneurial engineer.

10.2 Internal Funding

For start-ups, internal funding often begins with the entrepreneur's own assets, including time and energy. For established companies, new technology and product development is usually funded out of operating cash flow, although additional funds may also be sought to finance development of products for new markets. Industry sectors develop their own norms, over time, in terms of how much should be spent on these activities to remain competitive. Expressed as a percentage of sales revenue, expenditure on R&D can vary from 1% to 2% in industries with a low dependence on innovation up to 10–20% in the case of pharmaceuticals or the technology (IT) sectors. Companies that are not profitable enough to fund sufficient development end up being taken over or going out of business, inability to fund enough development being one of the classic causes of long-term business failure.

Companies will normally budget how much to spend on technology and product development, so specific programmes of work require, in effect, two rounds of internal approval:

- Agreeing overall R&D expenditure over, say, five years on multiple projects – normally a board decision on a rolling annual basis
- Agreeing the go-ahead for individual projects as they reach approval stages – normally delegated, depending on value.

Internal approval follows similar disciplines to external fundraising, as shown in Section 10.12.

10.3   Friends and Family Funding

If an individual, or a small group, are in the very early stages of developing a technology and are not part of a larger commercial organisation then, after personal resources, friends and family are realistically the only means of raising funds. These supporters will be investing as much in the person, or people, as in the technology itself. They will probably need some persuasion before parting with their money and hence a business plan, at least in verbal outline form, will be needed as well as the ability to convey the plan in a convincing manner, and answer questions. The latter will be good practice for future funding requests to banks or other independent funders.

Money from friends or family could come in the form of gifts, loans, or an equity stake (i.e. part ownership of the developing business). Whatever form is chosen, writing down and signing an agreement is always a good idea, as is a record of how the money is to be spent.

Family funders with some business experience may also play the role of advisers. They should be aware that engineering development is expensive when it comes to hardware, testing, and intellectual property protection, unlike, say, software development, where much can be achieved with little expenditure other than time.

They are likely to show some flexibility during the ups and downs of a fledgling organisation, provided that the founders show some sacrifice as well and communicate progress proactively. Funding of this type should hopefully take an organisation, and its ideas, to a point where it has progressed sufficiently to seek independent funding.

10.4 Angel Investors

As the next step after friends and family, 'angel investors' are one of the most frequent sources of seed funds for early-stage companies. Such investors are typically relatively wealthy individuals, or networks of individuals, who will provide funding from their own money for start-up companies in exchange for equity, or convertible bonds (loans that convert into equity), in the company. For tax reasons and to ensure that the entrepreneurs retain a majority stake, angels will not take more than 30% of a business's equity.

They are often experienced individuals who may invest in areas where they have expertise and will therefore also provide guidance and mentoring. Investment periods vary widely – some may fund for about 3 years, some may place a specific time horizon of their investment of, say, 5 years whilst others may fund for 10+ years.

The level of funding provided by angels varies widely but is typically in the low hundreds of thousands of pounds or dollars, although investments in excess of a million are known. Angel investors often pool their resources so that investments at a personal level may be in the tens of thousands. They frequently band together using 'clubs' attached to universities, business schools, or science parks.

An important point to note is that all early-stage investors will require a well-formed business plan, including a demonstration of business model, market understanding, and market risks. There will also obviously be technical risks, but these are usually more within the expertise of the start-up staff, so judging the character and experience of these people is an important but unseen part of the investors' due diligence. Entrepreneurs may be asked to 'pitch' to would-be investors, especially when they are acting in syndicate.

There are believed to be some 20 000 angel investors in the United Kingdom and a quarter of million in the United States, investing some £1.5 bn and $20 bn per year, respectively. See Refs. 10.1, 10.2.

In some cases, angel investments may be sufficient for the company to become viable, or sold on, but in many others funding from venture capital sources would then be sought to support the next stage of company development.

10.5   Venture Capital Funding

Venture capital (VC) funding is one of several forms of private equity ownership, i.e. where equity ownership of a business is not publicly quoted. It is typically used to cover the period from a firm's early-stage product and market development through to the point where that company is self-sustaining and capable of relying on public equity and/or bank debt (see, e.g. Ref. 10.3).

VC funds pool money from pension funds, insurance companies, private wealth funds, and sovereign wealth funds and manage it professionally. Fund managers actively participate in company management and decision-making, often bringing expertise in finance and marketing (but rarely in technology management). Securing VC funding requires comprehensive and credible business plans so that risks can be judged. It could be argued that VCs are more risk averse than angel investors, the latter coming in at a more formative, and hence less certain, stage of a company's development but putting in less money.

VC funding could typically be used for:

- Product development and intellectual property protection
- Marketing development and product launch
- Manufacturing start-up
- Working capital

VC funding may come in multiple rounds, known as Series A, B, C..., provided by the same investors or adding new ones as funding levels increase. Investments may be pooled amongst several VCs to spread risk.

Investments will typically be held for four to seven years and a return secured by an initial public offering (IPO) of shares, a trade sale or a sale to another VC fund. In the United Kingdom, £5 bn–£6 bn is typically invested each year by VCs in about 800 companies. The corresponding figures for the United States are $40 bn–$80 bn and 6000–10 000 deals per year.

Different VCs have different emphasis and strategy in terms of where in a company cycle they operate and may specialise in certain sectors, such as biotechnology or software. A current trend is towards somewhat later stage interventions, leaving reduced coverage for companies with technologies in the TRL 4–8 range.

10.6   Private Equity Funding

'Private equity' is a term that can be used generally to describe a situation where equity ownership of a business is not publicly quoted. It is also used more specifically to describe funds that acquire ownership or part-ownership of mature companies that are not fulfilling their potential. These might, for example, be divisions of large organisations, which no longer fit the corporation's strategy. The acquired organisation usually goes through a period of major shake-up and is then sold on, either publicly or privately, after a four- to seven-year period. This form of financing has little relevance to the main thrust of this book but is included for completeness.

10.7 Equity Crowd-Funding

A new but growing form of equity funding (and debt funding, sometimes known as mini-bonds or P2P lending) is through crowd-sourcing. The first examples of this form of fund-raising can be traced back to about 2010. Arguably, the concept is similar to the way charities operate – small contributions from a large number of private individuals. Unlike charities, the contributors each acquire a small equity stake in the company making the offer. When the figures are examined closely, though, it is obvious that a typical crowd-funding investment is more than just loose change.

The approach has been made possible by low-cost IT platforms, supported by social media, which are used to pitch the proposed investment. It is also the subject of financial regulation as the transactions qualify as the selling of securities as normally understood. So, for example, the Financial Conduct Authority in the United Kingdom and the Securities and Exchange Commission in the United States now accredit the operation of specific crowd-funding platforms.

The sums raised per offer vary widely, sometimes from hundreds of thousands of pounds (or dollars) up to more than $11 m for the largest example at the time of writing. Individual investments are in the low thousands, although professional investors may also take part and subscribe a lot more, sometimes taking the lead. The firms using crowd-funding come from every sector of business. For example, the author has had interaction with a mechanical engineering/renewable energy company that raised in excess of £2 m through this route.

In terms of where crowd-funding sits in the funding cycle, it seems to overlap with the later stages of angel investing and the earlier stages of venture capital funding. The further development of crowd-funding will be very dependent on how successful exits are from the first years of its operation, something that is largely unknown at the time of writing.

10.8 Bank Lending

Banks provide large quantities of business funding in a wide range of forms, typically as loans or overdraft facilities. Banks take a low-risk approach to lending, increasingly so since the 2008 crash. They will look particularly at the cash-generating capability of a business and hence its ability to support the payment of interest and repayment of the capital. They will also require some form of collateral, which could take the form of machinery, land, or buildings, all assets that can be sold in the case of business failure. Asset leasing and hire purchase are variations on the same theme.

Bank lending is therefore primarily of relevance to established businesses looking to finance ongoing operations or growth. Earlier-stage financing, if available, might have to be backed by personal assets, such as the family home.

There are also government-backed banks, such as the British Business Bank in the United Kingdom – see Ref. 10.4 – which operate as banks but back companies at an earlier stage of their development. Their loans are typically quite small, c. £25 k to $35 k, so they may have limited relevance as a primary source of funding for engineering development.

10.9   Peer-to-Peer (P2P) Lending

P2P lending follows similar principles to equity crowd-funding, using an IT platform to connect large numbers of small-scale lenders to borrowers, who might be private individuals or companies. The first examples emerged around 2005, and had mixed success with early problems of significant default rates by borrowers. The process is now more stable and robust. It is regulated by, for example, the Financial Conduct Authority in the United Kingdom and the Securities and Exchange Commission in the United States. The attraction of P2P is that it can potentially offer lower borrowing rates and higher savings rates than traditional banks. There has also been some government encouragement given the reluctance of traditional banks to lend to small companies.

In order to provide adequate confidence to lenders, the managers of P2P platforms have put in place criteria to assess companies before offering loans. These are the familiar criteria such as financial track record, accounting history and general creditworthiness. Hence, P2P lending is only likely to be feasible for established companies looking to grow and not feasible for early-stage companies.

10.10   Public Funding of Early-Stage Work

Public funding is widely available to support many forms of early-stage technology work, within the constraints of state aid legislation. In relation to the latter, the European Union, for example, defines (Ref. 10.5) three forms of R&D, which may be funded, at different 'intervention rates' (the percentage of a project's cost which a funding agency may contribute):

1) *Fundamental research.* Experimental or theoretical work undertaken primarily to acquire new knowledge of the underlying foundations of phenomena and observable facts, without any direct commercial application or use in view. Funding up to 100% of costs is allowed, but this category of work would precede the type of concept development work outlined above and is normally the province of universities.

2) *Industrial research.* Planned research or critical investigation aimed at the acquisition of new knowledge and skills for developing new products, processes or services or for bringing about a significant improvement in existing products, processes or services. This type of work might form part of early-stage concept work and is typically funded at a 50% intervention rate, usually requiring collaboration between two or more partners. Small and medium-sized enterprises (SMEs) are allowed higher levels than 50%, up to 75% in some cases. Work in this area is frequently described as 'collaborative R&D'.

3) *Experimental development.* Acquiring, combining, shaping, and using existing scientific, technological, business, and other relevant knowledge and skills with the aim of developing new or improved products, processes, or services. This may also include, for example, activities aiming at the conceptual definition, planning, and documentation of new products, processes, or services. A typical intervention rate is 25% and this is very much the territory of concept development.

There is a bewildering array of public schemes, operating within the framework above, which encourage and finance R&D work. They vary widely from country to country. Some are delivered by national bodies, such as Innovate UK in the United Kingdom. There are transnational programmes such as Horizon 2020 and its successor Framework Programme 9 (FP 9) in Europe. Then there are multiple regional and local schemes. All are based on the same premise: encouraging R&D leads to new technologies and hence products and services, which, in turn, leads ultimately to high-value jobs and economic growth.

Often, these bodies run periodic 'calls' where they request R&D proposals relating to a defined topic of interest with deadlines for submitting the proposal, which are then subject to independent review. Such calls are often oversubscribed, so there is competition for this funding.

Proposals are typically expected to cover:

- Why the proposed research topic has merit and (useful) novelty
- How it might be brought to market as a new product or service
- Who will participate in the programme and how it will be managed
- Why public funding should be used to support the work

Whether public funding is of interest to an organisation would depend on its preferences and circumstances. It may provide a welcome injection of funds for a tightly funded organisation. For others, the involvement of third parties, or some inflexibility to change direction quickly, may be unwelcome.

Public funding of early-stage developments will not cover all the costs of a new development, if only because intervention rates are less than 100%. However, public schemes are in active use by the smallest of SMEs through to major players such as Airbus, Ford Motor Company, and Shell.

The other route that governments use to encourage innovation is in the form of tax credits for R&D work. Different countries have different systems, but typically companies spending money to develop new products, processes, or services; or enhance existing ones, are eligible for a cash payment and a corporation tax reduction. In the United Kingdom, for example, up to 33% of eligible expenditure can be recovered in this way. The definition of R&D is quite broad:

> R&D for tax purposes takes place when a project seeks to achieve an advance in science or technology. The activities which directly contribute to achieving this advance in science or technology through the resolution of scientific or technological uncertainty are R&D.

Most importantly, the work does not have to be successful to qualify!

10.11   Public Development Facilities

Most major industrialised countries also have groups, or networks, of engineering development facilities, supported to a greater or lesser extent by public funding, and which can be used by engineering companies to undertake development work.

They concentrate on applied research, which can be taken up quickly by industry and therefore they are a step beyond university-style research, although there is overlap. The term 'research & technology organisation' (RTO) is used to describe them, and their European association, EARTO, estimates that their 350 members turn over some 23 billion euros p.a. in Europe. Their topics of interest are quite wide; not just engineering but biotechnology, digital technologies, semiconductors, energy, future cities, and healthcare are covered.

The largest network is within Germany where the Fraunhofer Gesellschaft has, at the time of writing, 69 institutes employing 24 000 people. Other facilities in Europe include the Catapult network within the United Kingdom, RISE in Sweden, VTT in Finland, and FEDIT in Spain. They are generally constituted as nonprofit organisations and are required to allow access to all comers on a nonpreferential basis. Their income derives from a mixture of public and private funding and they are active participants in European collaborative R&D programmes. Many have access to national funds set up to encourage technology development within small companies. For larger companies, the attraction is access to major facilities, such as wind tunnels, which single companies might struggle to afford.

In addition to large-scale engineering facilities, there are countless local schemes where business is encouraged through incubators and innovation hubs where office and workshop space is provided on competitive and flexible terms to early-stage companies.

10.12   Business Plans

External fundraising at whatever stage and in whatever form needs credible business plans. Without going into too much detail, these should be built around six points:

1) *Strategy.* Overall company aims, vision, and purpose
2) *Market.* Basis of competition, marketing plan, demand for the products or services, current and future client base, pricing, commercial model, volumes, territories, and competitors
3) *Product or service.* Details of current and future offerings, status of their development, development plans and timescales, risk analysis
4) *Management capability.* Details of key individuals, their capacity, responsibilities, skills, and track record of company leadership
5) *Financial plans.* Cash and financial projections as a function of time, financial track record, sensitivities, creditworthiness, accounts payable, and receivables history
6) *Exit route.* How equity investors will be able to exit (profitably) within a defined timescale

Internal business plans will follow a similar pattern with one or two omissions. See Chapter 5, Section 5.14, 'Approval and Formal Monitoring of Large Projects'.

10.13   Concluding Points

Technology and product development work has to be funded to make progress. Engineering R&D can be quite expensive with the need for computing facilities, test samples, and test facilities, as well as people's time. Money spent on work of this type is a form of investment, spent now with the hope of later benefits. These returns, unfortunately, are rather uncertain and could vary on any given project from zero to astronomical – hence, portfolios of investments are better from the point of view of spreading risk. All requests for funding need to be accompanied by a strong case, not just the idea but its business prospects.

Funding is available in multiple forms from the developer's own money through to private and public sources, all requiring the developer to have some of his own 'skin in the game'. William Shakespeare had his own views about financing:

> Neither a borrower nor a lender be,
> For loan oft loses both itself and friend,
> And borrowing dulls the edge of husbandry

His words are wise counsel when it comes to domestic finances, but business would quickly grind to a halt without borrowing and lending, especially in the field of technology and product development.

References

Four references from trade associations and a government-backed bank provide useful guidance, plus some facts and figures about various forms of funding provision:

10.1 Wright, M., Hart, M., and Fu, K. (2015). *A Nation of Angels – Assessing the Impact of Angel Investing across the UK*. UK Business Angels Association.

10.2 Pitch Book - (US) National Venture Capital Association – www.nvca.org

10.3 The Innovation Nation – UK Private Equity & Venture Capital Association, 2015, www.bvca.co.uk

10.4 (2015). *The Business Finance Guide - a Journey from Start-up to Growth*. British Business Bank, ©Institute of Chartered Accountants in England and Wales (ICAEW).

For those more deeply involved in European projects, or projects funded by European governments, this 38-page legal document is a comprehensive guide, albeit somewhat difficult to read:

10.5 Framework for state aid for research and development and innovation – European Commission, Brussels 21.05.2014 { SWD(2014) 163}, {SWD(2014) 164}

11

Running Teams and Working with Partners

11.1   Introduction

The emphasis of the previous chapters has been on the processes for developing new technologies or new products and it has covered such topics as:

- Engineering as a process
- Assessing the maturity of technology
- Aligning product development work with business and manufacturing strategy
- Planning and managing work
- Developing new concepts
- Identifying and managing risks
- Validating new products
- Delivering the final product information
- Obtaining funding

It is tempting to think of these processes as being largely mechanistic and therefore capable of being 'installed' in an organisation, as if they were pieces of software. The reality is that they are no more than guidelines or frameworks within which to operate. As has been said previously, technology and product development are iterative, learning activities where there are no black-and-white solutions. There is always some uncertainty about the best way forward, compounded by the fact that choices may have to be made with imperfect information.

On a day-to-day basis, the work will be handled by creative but fallible people, and the best processes are only as good as their interpretation by the people leading and running them. The work will also invariably be a team effort. The early stages of a new innovation may require a handful of people, no more than half-a-dozen perhaps. At the other extreme, a new aircraft with several million parts will clearly require thousands of people, who will be spread around the

Managing Technology and Product Development Programmes: A Framework for Success,
First Edition. Peter Flinn.
© 2019 John Wiley & Sons Ltd. Published 2019 by John Wiley & Sons Ltd.

globe. Either way, collaboration and cooperation are integral to the process, which is therefore subject to the vagaries of human behaviour.

This chapter is concerned, then, with human dynamics in the context of innovation, technology development, and product development. Most of the topics covered here are the subjects of lengthy books and have been researched in depth by social scientists. This chapter can give no more than a superficial overview of what might be relevant to the development of new technology and products. It is a rather important topic and one whose importance can be underestimated, as the wrong human dynamics can completely de-rail what might otherwise be a successful endeavour.

11.2   Working Collaboratively

The team, rather than the individual, is the basic building block of technology and product development – very little can be achieved by one person working on his or her own. It might be just one small, self-contained team developing a new idea or it might be a team, which is one of many on a much larger international project. The team is likely to be a mixture of full- and part-time members and is likely to comprise a core and then an extended range of participants, when all potential contributors are taken into account. Team dynamics will be a key determinant of the effectiveness of that team. So what does drive the success or otherwise of a team?

Starting on a slightly negative note, the work of the American writer Patrick Lencioni in *The Five Dysfunctions of a Team* (Ref. 11.1) has identified where things can typically go wrong in a team environment. In particular, he showed how the team can descend into a nonproductive, or at worst destructive, cycle of behaviours where one problem builds on another. The five areas he draws attention to are:

1) *Absence of trust.* Resulting in an unwillingness to be open and honest in the group
2) *Fear of conflict.* Ducking the confrontation of real issues, in an attempt to preserve an artificial sense of harmony
3) *Lack of commitment.* Half-hearted support of the group's work, resulting from the above
4) *Avoidance of accountability.* Not calling team members to account for non-productive behaviours
5) *Inattention to results.* Paying more attention to position of the individual and the avoidance of blame, rather than the success of the group

The author's experience of positive attempts at team-building, where, for example, a facilitator is employed to help a team develop and become more effective, places the focus on the opposite of these points:

1) Developing trust between the individuals and emphasising that team success brings personal success
2) Clarifying the roles of team members
3) Building on the strengths of team members (see also below under 'Team Composition')
4) Agreeing an overall purpose, strategy and approach to the team's work
5) Setting goals or tackling problems
6) Encouraging team members to bring issues out into the open but doing so in a way that is constructive and not seen as attacking other team members
7) Helping team members who are having difficulty with issues
8) Defining what constitutes success for the team and raising the commitment to get there

Motivated individuals can definitely be led towards better performance through these means and under-performing teams can be helped to overcome their difficulties.

11.3   Team Composition

Team effectiveness has also been very much helped by the work of Meredith Belbin, derived from his observations of the performance of teams working on management projects at Henley Management College. His work is used widely in the United Kingdom as a basis for developing team performance.

He observed that effective teams contained a balanced mixture of a number of different roles. The individuals on the team had a propensity to adopt those roles naturally (and one individual could adopt several of them). The nine roles now identified are:

1) *Plant.* A creator of new ideas and thus a useful contributor in the early stages of projects but less so in the latter stages.
2) *Resource investigator.* The investigator is good for networking and pursuing contacts and opportunities outside the team.
3) *Coordinator.* Originally named 'chairman', a coordinator is good at maintaining a balanced overview and helping decision-making.
4) *Shaper.* A shaper provides energy and drive, particularly when obstacles appear, and is prepared to challenge the team and shake it out of any complacency.
5) *Monitor evaluator.* These are relatively detached individuals who can see the facts of a situation and come to logical conclusions about the best way forward
6) *Team-worker.* As the name implies, team-workers are concerned with the smooth running of the team and have diplomatic skills that can defuse tricky situations.

	Technology Research	Technology Development	Product Development
Plant	✓✓	✓	
Resource investigator	✓✓	✓✓	
Coordinator	✓	✓	✓✓
Shaper		✓	✓✓
Monitor-evaluator		✓	✓✓
Teamworker	✓		✓✓
Implementer	✓	✓	✓✓
Completer-finisher		✓	✓✓
Specialist	✓✓	✓✓	✓✓

Figure 11.1 Team Roles versus Project Phase. Key: ✓✓ strong role; ✓ more limited role.

7) *Implementer*. This individual gets things done on time and can be relied on to tackle even the most difficult of tasks.
8) *Completer finisher*. The completer finisher ties up the loose ends and makes sure that tasks are undertaken properly to the required standard.
9) *Specialist*. As the name implies, specialists bring detailed, expert knowledge, within a defined field, to a project and are prepared to acquire new knowledge from the project, but again within their defined field.

Each of the roles above also has known downsides. For example, 'plants' often come up with new ideas in the late stages of projects when all is done and dusted; 'shapers' can become rather aggressive; and 'completer-finishers' can be irritatingly obsessed with detail. The team needs also depend on the stage of development of a specific project. For example, early-stage work benefits from plants whilst latter-stage work needs completer-finishers. Figure 11.1 suggests what roles are required as a function of project phase.

Whilst no framework is perfect, the Belbin structure does provide a basis for understanding and discussion. There is some academic criticism of the structure and the work behind it. However, it is a fact that it has helped many teams to understand themselves and to gain a practical understanding of how they may improve their functioning in a work environment.

11.4 Team Development

Even with the foregoing knowledge, teams do not suddenly come into being in an effective state; they need to work together to become so. Many people are familiar with the 'forming-storming-norming-performing' model developed by

Bruce Tuckman as long ago as 1965 (Ref. 11.2). This model does seem to replicate the states that teams in real life have to go through to perform effectively. These states follow a pattern that appeals to common-sense and experience:

- *Forming.* The team begins to understand the task and the team members, where people might fit in, where they might contribute, and something of the likely team dynamics.
- *Storming.* This concerns, as the name implies, a potentially difficult phase where conflicts and misunderstandings can arise, usually about individual roles or contributions and the extent to which they are acceptable to people as individuals. Sometimes these conflicts aren't fully resolved, bubbling away under the surface only to reappear later.
- *Norming.* The members of the group accept each other and their roles, perhaps understanding better the strengths of their colleagues and the collective strength of the group; there is at least a hint that the task can be achieved.
- *Performing.* The team gets on with the job with a sense of harmony and collective purpose, confronting issues in a constructive manner and feeling that results are being achieved, instilling a sense of pride in membership of the team.

Teams can, of course, become stuck in their early stages, never progressing, for example, beyond the tensions of the storming stage. Or, they can regress from the latter two states as a result of problems arising and disagreements over the way forward. A background awareness of this model is useful in guiding a newly formed team into becoming an effective unit.

11.5   Working with Partners

Almost without exception, engineering products consist of a mixture of in-house and externally sourced components. The proportion will vary, depending on the nature of the product. Those made from relatively basic materials may purchase 20–30% of the value externally and create the rest of the value in-house. More complex products such as aircraft or trains may buy 70% or more of the value and concentrate on system integration and final assembly, as well as post-delivery services.

The general trend is towards outsourcing an increasing proportion of a product's value, driven by increasing specialisation and capital intensity in the supply chain. Contrast, for example, Apple's current outsourcing of most of its iPhone manufacturing with Ford's famous River Rouge plant in Michigan, built between 1917 and 1928, which was set up to process iron ore at one end of the site and produce finished vehicles at the other. Whilst outsourcing may be presented as a relatively new concept, in reality, it is a continuation of the principles of specialisation, division of labour and comparative advantage identified some

CATEGORY	EXAMPLES	FORM OF COMMERCIAL RELATIONSHIP
Basic raw materials	Metal sheet and bar, plastic pellets	Transactional
Processed raw materials	Castings, forgings	
Commodity products	Fasteners, basic electronic components	
Make-to-print	Machined items, printed circuit boards, near net-shape items	
Proprietary items	Headlamps, switches...	
Proprietary systems	Aircraft engines	
Final assembly	Cell phone	Inter-dependent

Figure 11.2 External sourcing.

200 years ago by the economists Adam Smith (1723–1790) and David Ricardo (1772–1823).

The point is that engineering companies place an increasing reliance on external suppliers whose efforts need to be integrated into the wider product development teams. Their contribution will vary, depending on the nature of the material they are providing. In some instances, it may just be basic raw material with limited design input. In other instances, there could be substantial design input. Different categories of supply, and the form of relationship that is implied, are indicated in Figure 11.2.

The issues and principles discussed earlier in the context of internal teams still apply but are brought into sharper focus by the legal, commercial, and organisational barriers that will exist with separate organisations.

However, there is nothing to stop external suppliers being built into teams in the ways described above. In particular, the points described under 'positive attempts at team-building' are relevant: developing trust, clarifying roles, building on strengths, agreeing overall purpose, setting goals, encouraging team members, helping team members having difficulties, defining what constitutes success, and raising the commitment.

Most team-building or development seems to be internally focussed, but the principles could and should be applied elsewhere.

11.6 Working Internationally

A further dimension comes into play when cooperating organisations or individuals are of different nationalities. Unexpected social, behavioural, and commercial norms can create confusion and misunderstanding where none is intended and where everyone is genuinely seeking cooperation. It is therefore helpful to be, at least, aware of the issues that could arise in this situation and seek to avoid them.

The seminal work in this field (Ref. 11.3) is by Geert Hofstede, a Dutch engineer and social psychologist. In his more recent work, he identifies six dimensions of international culture, and he has profiled some 70 countries against these dimensions. The six are outlined in Figure 11.3.

The country profiles show that the United Kingdom and United States score highly in terms of individualism, masculinity, and indulgence but have low scores for power distance and uncertainty avoidance. France, on the other

Low Score	Dimension	High Score
Collectivism applies and people expect society to look after them in exchange for their loyalty.	**Individualism**	Individual choices and decisions predominate, people are expected to look after themselves.
Society believes that inequalities should be minimised and people should be treated as equals.	**Power distance**	There is acceptance that power in society is distributed unequally.
There is less competition and there is sympathy for the underdog, genders are emotionally closer.	**Masculinity**	Toughness, competition, and achievement are the principal drivers. Big is beautiful.
People are tolerant of uncertainty, happy to 'make it up as it goes along'.	**Uncertainty avoidance**	People feel threatened by ambiguity and therefore avoid uncertainty.
The emphasis is on the here and now, and the preservation of time-honoured traditions.	**Long-term orientation**	Preparation for the future is needed and built into society, perhaps through education.
Restrained culture, life is hard and duty is important.	**Indulgence**	Life is free and to be enjoyed with optimism and a positive attitude.

Figure 11.3 Dimensions of international cultures.

hand, scores somewhat lower on individualism (but still relatively high), masculinity, and indulgence, but higher on power distance and uncertainty avoidance. Germany is lower than the United Kingdom and United States on individualism and indulgence but higher on uncertainty avoidance and long-term orientation.

To an extent, these results correspond to national stereotypes, but they do provide some useful pointers in terms of what different nationalities might see as important when planning and running technology programmes. Some caution has to be exercised in that the data refer to nations as a whole, whilst particular individuals may have characteristics that depart from the national norm.

This work has now been extended and, to a degree, reinterpreted by Professor Erin Meyer of INSEAD. Her work (Ref. 11.4) is based on eight separate dimensions of national culture and is typically plotted in a way that enables easy visual comparison of the characteristics of different nations.

11.7   Working Virtually

The use of 'virtual' teams is becoming more widespread in a number of disciplines, including software and engineering, as is the topic of leading and managing teams in a cross-border, multicultural environment. Business school papers and personal development courses are readily available in both subjects. Arguably, engineering teams have always had elements of both, deriving in particular from the contributions made by international suppliers to large projects. A possible exception is small, early-stage teams, which tend to be single, co-located entities relying heavily on personal interaction to develop and refine new ideas.

The idea of virtual teams, where a large proportion of effort is located away from the central, co-ordinating group, has been made possible by two developments:

1) Low-cost communications technology such as email and teleconferencing which overcome, in part, lack of team co-location
2) Central, master databases of engineering information available in real time and capable of rapid updating so everyone can use current data

The potential advantages of this approach include the possibility of access to a wider pool of experts, lower development costs, and round-the-clock working. Set against these advantages are the reduction or loss of face-to-face discussion and the rapid iteration of new ideas that are typical of early-stage projects.

Research (Ref. 11.5) has suggested that, to make virtual teams work depends on the following:

- Having very clear definitions of the work required and the roles expected of those undertaking the work
- Investing effort in building trust between a project's participants, including some face-to-face meetings
- Having clear decision-making processes that respect the cultures of the participants
- Practising the skills of effective teleconferencing, backed up by frequent one-to-one support by the project leaders

These points suggest that virtual teams are most effective in the later, more detailed stages of engineering programmes when the direction has been set and there is more opportunity to subdivide interdependent tasks.

However, situations are never entirely black-and-white with virtual teams being one extreme of collaborative working, involving a high level of dispersion of activity, versus other projects where the bulk of participants are located closely but where there will inevitably be partners or suppliers on different premises. These points underpin the challenges of leading technology and product development projects.

11.8 Leadership of Technology and Product Development Projects

The leadership of technology and product development projects is a surprisingly multifacetted but potentially very satisfying and rewarding role. Most importantly, the leader is the holder and champion of the high-level vision for the new development; he or she is the 'master architect'. This in itself needs the person to operate with conviction and determination so that the team, small or large, can work towards the end goal with that same conviction. This aspect of the role needs a high level of personal integrity.

At the working level, the leader must be able to give guidance to junior staff and to specialists, who might be highly experienced but rather prickly characters who will probably believe that the leader lacks proper, in-depth technical knowledge. This aspect of the role needs good overall knowledge and understanding of the technology and its application. It is particularly important when problems arise, when solutions are needed, and therefore when choices must be made. Finding solutions is not easy, and they often emerge from debate, which the leader must be able to encourage. The role is very much one of an internal integrator, able to exert influence across a range of disciplines, always with the end goal and high-level vision in mind. It requires some humility and professionalism; conversely, it is not one that suits large egos.

Planning, coordination, and delivery are also important so that results are achieved within the agreed timescales and costs. This requires discipline of thought, planning and action. At the same time, the leader must instil a culture of freedom and responsibility.

Developing people and choosing potential successors is a critical aspect of the leadership role.

From an external point of view, the leader acts as the link to the outside world and the end customer. The role is often that of the internal voice of the customer. He or she must therefore be able to put themselves in the customers' shoes to act as their advocate.

There is also a very important role outside the team but within the wider internal organisation. This may be necessary to maintain financial support, to ensure support for progress beyond early-stage development, and to ensure 'fit' with broader organisational plans. The credibility of the leader is important with this group, and ideally the leader should have had business, marketing, manufacturing, or profit-making responsibilities in the past, i.e. 'real-world' experience.

The role requires a proactive, out-and-about attitude, meeting people at their desks, reacting quickly, and being highly communicative. Red-flag issues arise with some frequency and they must be dealt with competently and quickly. Hence, the role is not one where the person concerned can hide in an office looking at paperwork and reports or sending emails, although some of that is required (outside normal hours!).

11.9   Personality Traits

A further consideration is the personality traits that might or might not be most suitable for work in the field of technology and product development. There is clearly no one ideal mix of personality characteristics, just as there is no single type of role – different roles will be better suited to different personality characteristics and there is a wide variety of roles in this field. In terms of describing personality, there are several models which can be used. For example, the Myers–Briggs model, with 16 different combinations of characteristics, is used quite widely. As a start point for considering this topic further, the five-factor model described in Figure 11.4 is a straightforward starting point. The model has been developed independently by a number of research groups over a long period and is relatively easy to understand.

In terms of the traits which best match the needs of technology roles, it is fair to say that a high-level of conscientiousness is needed for all tasks in this field. The importance of team work would suggest higher levels of agreeableness and lower levels of neuroticism would be appropriate, although some level of neuroticism might be helpful where safety is involved. Beyond that, there

Personality factor	Description	Typical characteristics	Opposite characteristics
1. Openness to experience	This describes the level of openness to new ideas or experiences, the level of intellectual curiosity and the willingness to try new things.	Full of ideas, quick to understand, creative, have their own views	Unimaginative, cautious, struggles with abstract ideas
2. Conscientiousness	This is the tendency to act methodically and reliably and to apply self-discipline.	Organised, meets deadlines, can be relied upon, well prepared, attends to the detail	Dis-organised, impulsive, untidy, misses deadlines
3. Extraversion	This describes the extent to which an individual is stimulated by engagement with the external world versus a tendency to be more inward looking and solitary.	Depends on interaction with other people, gregarious, outgoing, action-oriented, sociable	More reserved socially, reflective, need periods alone
4. Agreeableness	This concerns the ability to work well and harmoniously with others and showing concern for others.	Trusting, helpful, able to get on with others, optimistic, willing to compromise	Not interested in or showing concern for others, sceptical, uncooperative
5. Neuroticism	This is the tendency to have negative feelings, irritation and emotional instability.	Tends to be anxious, pessimistic, and vulnerable to stress	Tends to be stable, calm and relaxed most of the time

Figure 11.4 Five-factor model.

are places for those with high or low levels of extraversion and openness to experience.

The critical question, though, is the extent to which these traits might correlate with, or predict, job performance and this is still very much a matter of debate. However, some knowledge of personality is undoubtedly helpful in understanding how individuals and groups function in a work environment. It at least reduces the possibility of putting square pegs in round holes.

11.10 Selecting People

Selecting people is arguably the most important role of any engineering manager — it will determine the success or otherwise of a team or department. The effects of selection are long-lasting and are difficult to unwind.

Choosing people for technology and product development roles is, in principle, no different to selection in any other field. The starting point is clarity over the purpose of the roles to be filled and the broad attributes that candidates need to bring. Any role to be filled should be viewed from two angles: first, the basic description of the role, which may not change much over time, and second the nature of the assignment currently. The latter will depend very much on current circumstances and could vary from delivery of new technologies, building a team, improving existing products, to supporting new business models. Each of these requires different attributes and experience.

It should be appreciated, also, that there is a wider variety of roles than might be expected in the technology field, which is sometimes perceived as the domain of the introverted analyst. Whilst there is a place for deep analytical and technological skills, most engineering roles have more to them than that.

For example, this book has continued to emphasise the strong interconnection between technology work and business issues, on the basis that the only real purpose of technology development is to create something that can be sold as a product or service. Increasingly, therefore, engineers will require broader business, marketing, and manufacturing awareness. This translates into the need for interpersonal skills to facilitate work with customers, suppliers, and manufacturers. The same could be said of influencing skills in the sense that all developments have to be communicated and 'sold' to managers or investors. Leadership skills are a key need at the more senior levels, both to lead engineering teams and to influence, more widely, organisations, customers and investors.

It is not the intention of this book to describe how recruitment processes or interviews should be conducted. There are many excellent books already in this field. However, it is important to be clear about the needs of particular roles and how candidates should be judged against those needs.

In terms of the process for selecting people, there is clearly a requirement in all technology roles for a high level of numerical, verbal, and spatial reasoning for which a number of established psychometric test methods exist. It is always as well to confirm that candidates are sufficiently strong in this respect, although degree-level qualification in a technically related subject should give sufficient confidence.

As far as selection interviews are concerned, a straightforward and effective approach for a process, with some pitfalls, comes from a rather unlikely source. The approach is outlined by Daniel Kahneman, Nobel Prize winner in economics, in his book *Thinking, Fast and Slow* (Ref. 11.6), which is more about psychology than economics. He developed the approach originally for the Israeli Defence Force for assessing the suitability of potential new recruits (and he followed up the process by confirming whether the recruits had developed into good soldiers). The basis of the process is that answers to a modest number of factual questions, pursued with some rigour, is more effective than relying on the 'expert' judgement of the interviewer, which proved to be 'almost useless for predicting the future success of recruits'. Other studies have shown that interviews are poor predictors of job performance.

He advocates constructing a list of six characteristics that are relevant to the role in question. As an example, in the previous section, the characteristics of a successful leader of technology projects was discussed and reference was made, as it happens, to six key attributes or roles for a leader:

1) Master architect
2) Internal technical integrator
3) Planner
4) Voice of the external customer
5) Organisational champion
6) Proactive but not egotistic style

For each requirement, a series of factual questions should then be compiled and worked through in the interview to evaluate objectively, on a 1–5 scale, how that person measures up. (This will obviously have to be done discreetly – not just working overtly and mechanistically through a prepared list.)

Selection is then made on the basis of this objective information, which is intended to establish whether the strengths of candidates match the needs of the job. The overriding thought is that 'normal' interview techniques are too easily swayed by subjective factors – first impressions, halo effects, horns effects, general confidence, and coincidental connections between the candidate and the interviewer. Ideally, this approach should be supplemented by discussions with those who have worked with candidates before.

It is often remarked that firm opinions about candidates are formed in the first 20–30 seconds, on the basis of appearance, dress, tone of voice, or body language – see Ref. 11.7. Subsequent interactions then seek to confirm this early view. There is also a natural human tendency to generalise one particular success to overall success (halo effect), or the opposite of generalising one failure into overall underperformance (horns effect). The wider point is that human beings are naturally 'programmed' to make rapid judgements, and this works against an objective assessment of who will perform best in a job, which is the ultimate purpose of a selection process.

At the same time as judging the basic competence to undertake a role, an assessment is needed of the social skills of candidates and of their underlying values, which need to align with those of the organisation. The latter can be built into, or derived from, the core questions whilst the latter is something that will come across in an interview (social skill probably wasn't a major consideration in the military example quoted above!).

Competition is fierce for high-calibre candidates for technology and product development roles, and these points will help to identify the best candidates. Those candidates will also become aware of the process and be more likely to accept positions with organisations that have an obviously professional selection process.

11.11   Developing People

The core processes described in this book depend critically on the skill, judgement, and experience of the people leading and participating in those processes. These capabilities are not developed overnight. As has been said elsewhere, a full engineering cycle takes several years, at least, and may even extend to decades in some instances. Long-term internal development of staff is therefore likely to be more effective than a wholesale reliance on external recruitment.

The primary means of personal development is through the work itself, not through an endless succession of training courses. Given that young recruits into engineering will be well-armed with analytical and theoretical skills when they enter the profession, the aim should then be to broaden people's capabilities in areas such as programme management, solution design and practical development, customer/partner/supplier relationships, manufacturing introduction, and team leadership.

The speed of personal development can be accelerated by providing the right opportunities to acquire these skills, supported by guidance and mentoring. Finding the right opportunities does, of course, depend on the circumstances of each particular organisation. It also depends critically on the organisation to identify these opportunities and advocacy in putting people forward for work where they may not have the full set of skills. These are not issues where the individual can easily make his or her own way – the organisation should take the lead, and it is in its own interests to do so because the future of technology and product development organisations depends on its future talent.

There is a part to be played by formal training, but it is likely to be only a small role. Beyond 'on the job' development, exposure to the following can be helpful:

- International technical conferences where there is opportunity to meet like-minded people in related fields
- Training in specific technical topics

- Professional society activities, which also provide opportunities for meeting other professionals and may include the chance to visit other firms or countries
- Personal development workshops in the 'softer', human skills
- Business education in specific topics such as marketing or more widely up to MBA level

A broader question, then, is who should take responsibility for personal development. At one time, engineers could, if they wished, look forward to a lifetime's work with one company. In this case, the company would often take responsibility for that person's development. Companies now have much shorter lifespans, so responsibility therefore must rest with the individual. The individual would benefit, however, from professional help in areas such as personal strengths and values, where one can contribute most, assessing one's own performance and one's learning style.

In his book on this topic (Ref. 11.8), Peter Drucker makes the point:

> Now, most of us… will have to learn to manage ourselves. We will have to learn to develop ourselves. We will have to place ourselves where we can make the greatest contribution. And we will have to stay mentally alert and engaged during a 50-year working life.

11.12   Concluding Points

The main message of this chapter has been that human dynamics plays an important part in the running of engineering activities. This should come as no surprise, as these activities are often complex endeavours, undertaken by groups varying in size from a small handful to several thousands, using well-educated and opinionated engineers. Whilst classic, technical work might be the core of the activity itself, its successful execution requires organisation, as well as constructive interaction with customers, investors, and other stakeholders. Some understanding of human dynamics is therefore an essential part of the armoury of anyone in a leadership position in this field.

This chapter has introduced some models and ways of thinking that might be helpful in guiding how people are selected and developed, and how activities are organised. It provides no more than a practical introduction to a range of complex subjects, which are themselves deeply researched and where there is often controversy about what is right or wrong. As always, the technology leader has to take a pragmatic view of how to develop and use an understanding of these topics.

References

There is a plethora of literature on the social science of organisations of which this is just a small, but relevant sample. All are specifically referenced in the main text.

11.1 Lencioni, P.M. (2002). *The Five Dysfunctions of a Team*. San Francisco: Jossey-Bass/Wiley.

11.2 Tuckman, B.W. (1965). Developmental sequence in small groups. *Psychological Bulletin* 63: 384–399.

11.3 Hofstede, G., Hofstede, G.J., and Minkov, M. (2010). *Cultures and Organizations: Software of the Mind*. New York: McGraw Hill.

11.4 Meyer, E. (2014). *The Culture Map: Breaking through the Invisible Boundaries of Global Business*. New York: PublicAffairs Books.

11.5 Watkins, M.D. (2013). *Making Virtual Teams Work: Ten Basic Principles*. Boston: Harvard Business Review.

11.6 Kahneman, D. (2011). *Thinking, Fast and Slow*. New York: Penguin Books.

11.7 Dale, M. (2006). *The Essential Guide to Recruitment*. Kogan Page.

11.8 Drucker, P.F. (1999 and 2005). *Managing Oneself*. Boston: Harvard Business Press.

12

Decision-Making and Problem Solving

12.1 Introduction

As has been frequently stated in this book, the engineering development process revolves around iteration and learning. New facts are continually emerging, and therefore decisions are constantly being made. Problem solving and decision-making are, from one point of view, just everyday elements of technology development work. Some of these decisions will be minor and hardly be recognised as decisions at all. Others will be substantial, requiring careful thought and objective analysis before a conclusion is reached.

The ability to apply the mind well and solve problems is vital to work of this nature. The processes of 'critical thinking' are therefore very relevant.

However, reaching the right decision, if there is such a thing, may not be quite as straightforward as might be thought. As will be shown, the human decision-making process is subject to all sorts of biases and prejudices that get in the way of choosing the best way forward.

Research in the field of behavioural economics has provided some of the best insights into how the human mind works in this respect. This research has shown that human beings, when presented with 'rational' economic choices, behave far from rationally. The findings of this research apply across the spectrum of human endeavour, but particularly to technology work.

The other field that impinges on technology work is statistics. 'Statistical thinking' has revolutionised the approach to manufacturing over recent decades. Started in Japan under the guidance of the American, Dr W. Edwards Deming, thinking of manufacturing as a series of processes with their own characteristics expressed in statistical terms has resulted in a radical change in the management of those processes (Ref. 12.1). In particular, it has produced that elusive combination of better and faster results at lower costs. As a result, many companies now train their people in Six Sigma, probably without realising its true meaning.

Elements of the same statistical thought processes should be applied to technology and product development work. Of course, the quantities involved

Managing Technology and Product Development Programmes: A Framework for Success,
First Edition. Peter Flinn.
© 2019 John Wiley & Sons Ltd. Published 2019 by John Wiley & Sons Ltd.

in this work, often 'one-offs', are much lower than in manufacturing, which limits the extent to which detailed statistical analysis can be applied. However, this does not prevent the use of the principles of statistical thinking.

The field of decision-making is one where some level of awareness is beneficial to everyone involved in technology development. This chapter aims to provide a start point for study of this field. As with many of the topics covered by this book, there are much more detailed, and rigorous, publications on this subject.

12.2   Decisions to be Taken

As noted above, decisions permeate development processes. Decisions are required in many areas, such as:

- Choice of project to pursue
- Technical route to be followed
- Drawing conclusions from customer data
- Choice of solutions to problems encountered
- Type of analysis or test work to be undertaken
- Whether to continue to invest money
- Allocation of work among individual engineers
- Selection of people for employment

Some of these decisions will be taken in a few moments by one individual. Others may be the subject of detailed and documented analysis for review by a senior company committee. The point is that there are thousands of decisions, small and large, short-term, and long-term, to be dealt with in a development programme – all potentially subject to the vagaries of the human mind. This raises the question of how those decisions can best be made.

In principle, any significant decision should be made on the basis of a cost–benefit analysis. Costs can usually be calculated but benefits, which could be nonfinancial, may need more thought. However, the principle stands. Alongside costs and benefits, the opportunity costs should also be considered. All decisions come with an opportunity cost in terms of alternative uses for money, resources, or expertise, and hence the lost benefits of an alternative course of action. Rational evaluation of these points should underpin every decision.

12.3   Critical Thinking

The study of rational processes of human thought can be traced back more than two and a half thousand years to Greece and the times of Socrates, Plato, and

Aristotle. 'Critical thinking' seems to be a more modern term, although the word *critical* has Greek origins. In today's world, it is a recognised term relating to the objective and rational evaluation of an issue in order to draw a sound conclusion. The importance of the topic is reinforced by the existence, in the United States, of the US National Council for Excellence in Critical Thinking and, around the world, numerous other centres and hubs for this topic. In academic circles, the subject is taught quite widely and forms the basis of rigorous academic analysis.

Another US organisation, the Foundation for Critical Thinking, based in California, describes the proficient critical thinker as someone who:

- Raises vital questions and problems, formulating them clearly and precisely
- Gathers and assesses relevant information, using abstract ideas to interpret it effectively
- Comes to well-reasoned conclusions and solutions, testing them against relevant criteria and standards
- Thinks open-mindedly within alternative systems of thought, recognising and assessing, as needs be, their assumptions, implications, and practical consequences
- Communicates effectively with others in figuring out solutions to complex problems

To these points, it could be added that the proficient thinker is also aware of human fallibilities in reaching decisions and is competent in the use of statistical methods in arriving at sound conclusions.

These rather philosophical points underpin the following discussion about decision-making in the field of technology development.

12.4   System 1 and System 2 Thinking

In the work by Daniel Kahneman referenced in the previous chapter (Ref. 12.2), the concept of two distinct modes of human thinking is described, based on classifications developed by the psychologists Keith Stanovich and Richard West. These two modes are described, somewhat unoriginally, as System 1 and System 2.

System 1 looks after short-term reactions to situations as they develop. It is instinctive and fast. Some of its skills are 'built-in', such as responding to a loud bang or ducking a flying object. Other skills are acquired through learning but become instinctive: language, car driving, or playing musical instruments. System 1 reacts to incomplete information so it is fast but imperfect. It links to the choice of 'fight or flight'.

System 2 is slower but essentially rational. It relies on deeper thought, analysis, and reasoning. If applied, it can in effect override the instinctive reaction of System 1.

There is some academic criticism of this simple model; there are those who argue for example for four distinct systems and others who argue for a graduated transition from one mode of thinking to another. However, the two-system model does correspond with general, human experience, and it is useful to bear in mind when faced with a pressurised decision-making situation.

It might be argued that technology development work is entirely the domain of System 2. However, in the hurly-burly world of engineering delivery projects with tight timescales, hasty decisions, subsequently regretted, are not unknown.

An extension of this model concerns the unconscious mind (Ref. 12.3). We are all familiar with the experience of finding the solution to a knotty problem (at least the principles, if not the detail) after 'sleeping on it'. Studies of people who are regarded as very creative all report flashes of inspiration, almost from nowhere, relating to problems or topics that have often been under review for weeks and months. Psychologists have concluded that there is such a thing as the unconscious mind and that it acts in the background in a very powerful manner. It registers much more environmental information than the conscious mind can process and seems to be capable of solving problems that are beyond the reach of the conscious mind.

The relevance of these points lies in gaining greater understanding of the mental processes at work when attempting to manage technology development. This understanding may be helpful in achieving better results from the work.

12.5 Human Barriers to Decision-Making

In addition to the points made above, it needs to be noted that the human mind is susceptible to all sorts of biases and fallacies which may obstruct logical thought, especially on complex issues. Many of these stem from a human tendency to form quick conclusions based on imperfect evidence and a tendency to gross overgeneralisation based on a small amount of information. This weakness probably derives from ancient survival instincts. Some of those relevant to the work of this book are described below:

Confirmation bias. We have preconceived views on almost every topic we encounter. These preformed hypotheses tend, then, to bias the way we look at information, emphasising data that supports our view (perhaps subconsciously) and filtering out (again subconsciously) data that go against our view. An extension of this bias is conducting experiments that only test one hypothesis and don't therefore allow other conclusions to be reached.

Post hoc fallacy. Described fully as *post hoc ergo propter hoc* (translated as 'after this so because of this'), this logical fallacy is the erroneous connection of one event to a preceding event, especially if they are very close in time. Many of these fallacies are no more than superstitions: 'I always play better football if I tie my left laces first', for example. In engineering work, problems can arise in rapid succession, and it is important to understand which are related and which just happened to occur very closely in time.

Sunk cost fallacy. When a lot of money has been spent on a development, there is a natural reluctance to abandon it, even if the prognosis is that the project will not lead to a successful outcome. The decision about the future of a project should be based on future costs and revenues, independent of sunk costs. If the prognosis is poor, don't throw good money after bad and accept that the earlier investment was wasted.

Status quo bias. Another natural human tendency, when faced with a difficult choice, is to revert to the status quo, the security of what is already known, the safe option. It is linked to two other recognised biases: loss aversion (potential losses being more highly rated than potential gains of the same magnitude, generally by a factor of about 2) and endowment effect (a preference for keeping hold of what is already possessed). As the terms imply, the research in this area derives from behavioural economics. In technology work, it manifests itself in a reluctance to make changes when evidence is pointing to the need for change. The expression 'bite the bullet' comes to mind.

Attribution error. This well-known phenomenon concerns the way we judge people's behaviour, and is therefore relevant to successful teamwork or management of people. In essence, it is the tendency to judge other's behaviour as being a consequence of who they are, intrinsically, rather than its being a product of the situation in which they find themselves. Conversely, we judge or explain our own behaviour more as a reaction to circumstances and hence more favourably. A one-hour car drive, with the usual encounters with 'idiot drivers', will illustrate the point.

Reliance on second-hand information. This is not a recognised human fallacy, in the world of social psychology, but is an issue frequently experienced by the author. When discussing problems and making decisions, it is surprising how frequently and confidently people are prepared to form opinions without any first-hand experience. It is just as surprising to see the effect of experiencing issues first-hand and the changes it often brings about to those opinions. A useful dictum is to refuse to discuss any issue before experiencing it directly.

The six points described above can all come into play, not usually together, when making decisions. They can get in the way of making effective and often difficult decisions. Hence, when faced with a difficult decision, it is sometimes a good idea just to stand back and reflect on whether the thinking about that decision is being biased by any of these points.

12.6   East versus West

Whilst discussing mind-sets and decision-making, it is worth reflecting on the differing approaches taken in Eastern countries – Japan, China, and South Korea – to those in Western countries – Europe and the United States. In his book *The Geography of Thought* (Ref. 12.4), Richard Nisbett suggests there are fundamental differences in the way that East Asians think about issues and decisions, compared with Westerners. Anyone with experience of engineering business in these two cultures will know this to be the case, although those with this experience will also know that there are wide variations between:

- Western countries, e.g. United Kingdom, United States, Germany, France, and Italy
- Eastern countries, e.g. India, China, Japan, South Korea, and Malaysia
- Specific companies, with their individual cultures
- One person and another

Westerners are brought up and educated, whether they know it or not, on the basis of rational logic. Facts are gathered, logical truths follow and, if a contradiction occurs, further analysis will resolve the rights and wrongs of the contradiction. Once these points have been dealt with, the way is clear to proceed! So, if $A = B$, and $A = C$, then $B = C$. Hard logic of this type leaves no room for ambiguity. This approach is the basis of scientific theory, the development of which over the past 400 years has been undertaken largely in the Western world. It is interesting to note, however, that much of recent development in quantum mechanics, for example, is based on theories where more than one answer is possible, e.g. wave-particle duality, uncertainty principles and entanglement, and hence where classical logical principles appear to be overruled.

Eastern traditions, on the other hand, place much more emphasis on the environment or context surrounding the 'facts'. They are more willing to live with contradictions, to develop a point of view by dialogue (so-called dialectical reasoning), and to maintain some level of ambiguity as events proceed. Events inevitably bring change, and hence a new 'answer' becomes appropriate. There is also a greater willingness to explain human behaviour through the context in which an individual is operating, rather than through the intrinsic traits of that individual. Some Eastern countries also place much greater emphasis on data and use the gathering and discussion of data as a means of building consensus before taking action. In other Eastern countries, 'loss of face' is a major driver of behaviour.

It is interesting to consider the extent to which Eastern countries', especially Japan's, recent success in technology and product development has been based around this mind-set. It is certainly the case that many Western companies have changed their priorities in the last two decades, have taken a more process-centric approach and have thereby adopted many of the Japanese principles of product development. It would seem that the most successful organisations are likely to be those which can make balanced use of both the Eastern and Western traditions.

However, it is difficult to be entirely prescriptive on this topic, and the most important preoccupation for any engineer going into a new, international cooperation is to be aware that significant cultural differences exist from country to country, company to company, and individual to individual.

12.7   Statistical Thinking

A good definition of statistics is 'that branch of mathematics dealing with the collection, analysis, interpretation, presentation, and organization of data' (Ref. 12.5). The subject is usually associated with large quantities, or series, of data from which inferences can be drawn. It is certainly the case that technology and engineering work will generate data sets in various circumstances to which classic statistical methods can readily be applied. Examples might include:

- Data from undertaking customer surveys
- Time series information from data logging on a prototype machine
- Data from manufacturing operations over a period of time
- Warranty and complaint information from products in the field
- Increasingly, data from 'connected' products out in the field

In larger companies, engineers trained in statistical methods will be able to analyse such information and make good use of it.

However, it is also often the case that some data are available, but in very small quantities, and decisions still have to be made. Applying a 'statistical mind-set' often helps such decisions. Lawrence Summers, the sometimes controversial former president of Harvard University, stated: In an earlier era, almost every student entering a top college knew something of trigonometry. Today, a basic grounding in probability, statistics, and decision analysis makes far more sense' (Ref. 12.6). These are wise words.

First, consider that information or data can usually be grouped into 'populations' of related information. The five groupings noted above, which could contain substantial amounts of data, could be considered as populations. It might also be a small grouping, such as a number of separate R&D projects or a small number of candidates for a job.

Large populations are normally analysed or characterised through samples of data, which clearly have to be representative of the population as a whole. Populations can then be characterised by parameters such as the mean and standard deviation. In the engineering world, dispersion or wide variability in the data about the mean usually causes problems. For example, in a manufacturing process, there is a need to hit a dimension on a drawing accurately, within narrow tolerances. There will be some process variability but it should be minimal, and one of the constant objectives of manufacturing engineering work is to

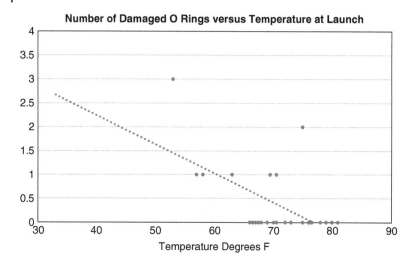

Figure 12.1 Challenger O-ring failure rate versus temperature with simple linear extrapolation (exponential extrapolation is worse).

reduce variability. This can be done by understanding the details of the process, including the causes of variability, and developing improvements.

Turning to small populations of data, analysis and presentation of key facts requires more thought. This was illustrated when the causes of the Space Shuttle *Challenger* disaster were analysed. The issue came down to the low-temperature performance of the O-ring seals in the joints between the different sections of the booster rockets. There had been some failures on previous missions. Putting aside all the issues of management culture exposed by the enquiry, and with the benefit of hindsight, a simple presentation (and a willingness to take it seriously) of failure rate versus ambient temperature (Figure 12.1) might have clarified that a serious problem existed.

This is an unfortunate but clear example of a situation where an understanding of data and statistical thinking would be beneficial to everyone.

A more positive, and older, example of clear presentation of data can be found in the work on the causes of cholera, undertaken in the nineteenth century by Dr. John Snow. At that time, cholera and a number of other diseases were thought to be caused by the 'miasma', some sort of mysterious heavy vapour. John Snow had first experienced a cholera outbreak in 1832. After an outbreak in London in 1849, he published *On the Mode of Communication of Cholera* proposing that cholera was caused by contaminated drinking water, but his theory was not accepted. He then studied a further outbreak in 1854 centred on Soho in London in which over 600 people died. He plotted the exact location and numbers of the deaths on a map (Figure 12.2). In 1855, he published a second, much expanded edition of his work.

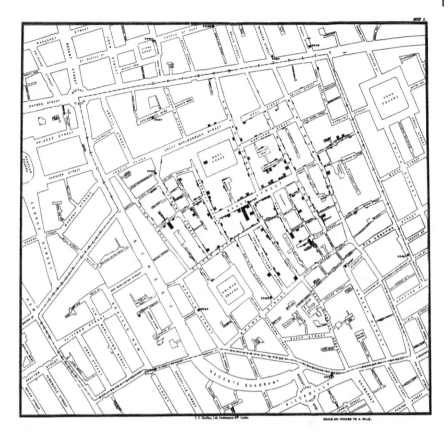

Figure 12.2 Map by John Snow showing the cholera cases in the 1854 London epidemic. Source: Re-drawn by Charles Cheffins.

Close inspection of the nineteenth-century plot shows quite clearly that the outbreak centred on a water pump at the junction of Broad Street and Cambridge Street. Snow's further and careful analysis also explained why some nearby places had no outbreaks (the brewery, for example, which had its own water supply) and someone in Islington suffered (she liked the taste of the Broad Street Water and had it ferried in). A twenty-first century replot of the data with colours and various sized dots shows an even clearer picture.

More examples of how best to present numerical data are given in Ref. 12.7.

12.8   Application to Management Processes

Some of the same principles could be applied to management processes. Take recruitment as an example. Studies have shown a 0.1 correlation between interview results and success in the job, i.e. the two are only very weakly related. This is hardly surprising – the sample here is a one-hour interview versus, as an alternative, data samples from the real work a person may have undertaken over many years. If an organisation's recruitment process produces highly variable results in terms of the effectiveness of new recruits, then place more emphasis on finding out how effective candidates have been in their previous work and less emphasis on interviews.

As a slightly humorous example to illustrate how the wrong sample can mislead, the author has had one 20-minute sample of the golfing prowess of the Japanese golfer Tsuneyuki ('Tommy') Nakajima. It happened to be at the 17th hole in the third round of the Open Championship at St. Andrews in 1978. It took him nine shots to sink the ball into the hole after having some trouble in a well-known bunker. Based on that small sample, he would have trouble getting into a game with a normal Saturday morning fourball, let alone entry to professional competitions. He was, of course, one of the most successful Japanese golfers ever.

This example can also be used as an illustration of the concept of 'regression to the mean'. This concept states simply that, after one extreme measurement, the next measurement is likely to be much closer to the mean. Tommy confirmed this principle by doing much better on the 18th hole. The point is that extreme results can occur even in stable processes and there should be no overreaction until there is more evidence that a process has radically changed.

Another example of a typical engineering management decision, where previous data should be taken into account, concerns product introduction timing. There is always pressure to speed up the process but it is worth reflecting on how long previous projects have actually taken (not what they should have taken) from approval to actual introduction – including an estimate of the mean and standard deviation, even if there are relatively few data points. If the proposed introduction timing is radically faster than has been achieved previously, and especially if there is no strategy to achieve it other than 'trying harder', then a statistical review of the proposal might suggest a different decision.

The overriding point of the sections above is that all decisions are based on information and, usually, numerical data. It is very helpful to frame information, data, and decisions in a way that is consistent with statistical principles and to use these principles to challenge the potential solutions being put forward.

12.9 Problem Solving – A3 Method

Problem solving and decision-making are closely linked. Fortunately, there are some very straightforward methods of helping the process of solving problems. A quick scan of the internet will reveal numerous methodologies, usually designed around a certain number of steps, where that number can be anything between 4 and 10.

All methods follow a similar pattern of identifying the problem, characterising it, coming up with solutions, trying them out, and confirming a workable solution. This is essentially a slightly longer version of the iterative process outlined in Chapter 2.

A good method, which is widely used and may have its origins in Toyota, is based on an A3 sheet of paper – – see Ref. 12.8 and Figure 12.3 for the format. This approach makes it easy to display the work in, for example, a team room so others can see what is happening and make a contribution. Shown in Figure 12.3, there are nine key steps:

1) What has caused a problem to be identified and recognised, i.e. what are the symptoms of the problem? These may be described in nontechnical or semi-technical terms, such as 'the bracket holding the fluid reservoir has broken'.

2) How is the problem defined? This may require some further observation, data gathering, and analysis; the more thorough and technical the definition, the better the problem can be understood. An example might be: 'the fluid reservoir mounting bracket has failed through fatigue cracking between two bolt holes after 10^5 cycles of the endurance test to test code xxx'.

3) What should the situation be when the problem has been overcome? Again, good definition will help derive a solution and will give a point of reference against which to test solutions; for example: 'the bracket should withstand 10^6 cycles of the standard test'.

4) What root causes can be established as the source of the problem? Arguably, this is the most important step; if the root cause(s) can be established accurately, a well-designed solution is far more likely. In the example above, the root cause might be a manufacturing method that introduces surface defects from which cracks can develop and propagate.

5) What solution(s) can be proposed? There may be several; in fact, a choice of solutions is ideal. Solutions to the above problem might include different manufacturing methods, or some form of surface treatment, or redesign to spread the loads more evenly.

6) How are the solution(s) to be taken forward, in terms of review, analysis, testing, and implementation? The different solutions might be costed and analysed.

7) What results have been obtained from the development of the solutions, and what choice should therefore be made?

8) How has the chosen solution been shown to be effective in meeting the requirements defined in (3)?

9) What lessons have been learnt for future use?

A3 PROBLEM SOLVING SHEET

Team members:		Team Lead:

1. Symptoms Observed	4. Analysis of Root Cause	7. Results from Action Plan
2. Problem Definition	5. Proposed Solution	8. Confirmation of Solution
3. Target Result	6. Action Plan	9. Lessons for Future

Date:	Reference:	Author:

Figure 12.3 One-page A3 problem-solving model.

This process does not guarantee a solution, but it does provide a sensible and open structure for dealing with problems. By making the process explicit, it helps other people, including those with more experience, to make a contribution. The focus is on a clear definition of a problem and an organised approach to solving it.

The example above relates to a development test and therefore a limited number of test samples, possibly only one. If it was a problem in service, the approach would be somewhat different and would have to take into account a wider population of components. Further questions might include:

- The number of components in service
- The failure rate amongst those components
- Whether the failure occurs uniformly across the population or whether it is related to a particular batch
- Whether the failure is linked to a specific user or duty cycle

There are also supplementary methodologies that can be used to support some of these stages. For example, the '5 whys' method is a very good way of moving from symptoms to root causes. This method, very simply, consists of successively asking the question 'why' to unearth the root cause of a problem. Asking the question five times usually works, although it may require anything between 3 and 10 cycles. For example, in the instance above:

1) Why did the bracket fail? – fatigue cracking
2) Why did fatigue cracks occur? – surface imperfections
3) What caused the surface imperfections? – insufficient clamping force during machining operations
4) Why did the clamping problems occur? – lack of previous experience of this type of operation
5) Why was there a lack of experience? – peer review of machining operation was cut short

This example is somewhat trite but it does illustrate the process. There is some criticism that '5 whys' can be oversimplistic and one-dimensional, particularly when problems are more complex, but it is a good starting point for thinking about the causes of problems. It focuses attention on the underlying causes of problems and not just the symptoms. Although it might not provide a complete answer, it is always worth thinking through before undertaking a more complex analysis.

It is also important to appreciate that some problems are relatively straightforward whereas others are more intractable. Straightforward issues can often be solved by a single individual following a logical and linear process, working through the calculations to a single solution. A lot of engineering development falls into this category. Other problems may be more difficult to resolve within the constraints of performance requirements, material properties, space availability, procedural requirements, cost, and manufacturability. These latter will require a range of ideas from different people and may result in a range of solutions from which the best compromise is chosen.

12.10   Creative Problem Solving – TRIZ Method

The method just described relates to well-defined, or bounded, problems to which a specific answer is sought or where a specific decision is required. When it comes to more creative work, where a creative engineering or technological solution is required, the TRIZ method (Ref. 12.9) has been widely taught and used. It is aimed at finding innovative solutions to problems where the root cause is largely understood. Rather than relying just on the instinctive creative talents of the engineer or scientist working on a problem, the method is intended to give structure to an otherwise unstructured process.

TRIZ has its origins in Soviet Russia, where it was first developed, after the Second World War, by Genrich S. Altshuller. Despite spending time in a labour camp during the Stalin era, he persisted in developing a methodical approach to solving technical problems, based on his thousands of observations of inventions described in patents. (Like Albert Einstein, he worked in a patent office for a period.) The method did not emerge from the former Soviet Union until the 1990s, since when it has developed a significant following and is taught quite widely.

In Russian, the acronym TRIZ stands for *Teoriya Resheniya Izobreatatelskikh Zadach*, which roughly translates as 'Theory of Inventive Problem Solving'. It has a focus on problems that involve contradictions or trade-offs. For example, making something stronger could involve also making it heavier – so how can the strength be improved without a corresponding weight increase?

The method has been criticised as being difficult to follow; there are, about 40 different solution types that can be used, but it is not clear which should be used in a particular situation. However, there are route maps to guide the process, and the reference by Gordon is particularly good in this respect (Ref. 12.10).

Altshuller's original work identified five different levels of problem difficulty and solution sources. Level 1 is the simplest and generally used solutions are already known in the industry concerned. Level 3 problems used known solutions but from another industry whilst Level 5 required new scientific discoveries, e.g. solid-state semiconductors in place of thermionic valves.

The process advocated by Campbell starts with an analysis of the problem via the 'classic' route outlined in the previous section, with its emphasis on identifying and structuring the problem. It only then proceeds along the TRIZ route if a solution cannot be found by standard methods. Summarised in Figure 12.4,

STAGE	APPROACH	DETAILS	
1. Identify Problem	Standard problem-solving method, leading to need for TRIZ process or otherwise		
2. Select Problem Type	Specific problem	Cause unknown	
	Specific problem	Cause known	
	Improvement/ development		
	Failure prevention		
3. Apply Analytical Tools	Root cause analysis	Root cause analysis incorporating, TRIZ tools algorithm, CEC-1, SA-1	
	Analytical tools	Ideal solution system, CEC-2, Functional model & trimming	
	Analytical tools	CEC-3, S-curve analysis, Trends of evolution, Anti-system	
	Sub-version analysis	Subversion analysis, Sa-2, CEC-4	
4. Define Specific Problem	Define specific problem(s), including contradiction statement		
5. Apply TRIZ Solution Tools	Model of Problem	Model of Solution	
	Technical contradiction	40 inventive principles	
		Separation, satisfaction, bypass	
	Physical contradiction	76 standard inventive solutions	
	Su-field model	Scientific effects	
	Function statement	Trend of evolution	
	Search for trend solution		
6. Solutions & Implementation	Specific ideas and solutions leading to implementation		

Figure 12.4 TRIZICS solution roadmap.

the first step in the TRIZ route is to assign the problem to one of four different problem classes.

There is then a series of standard TRIZ tools for analysing the problem according to its type. Specific problems are then formulated from this analysis, often in the form of a 'contradiction' statement. The classic TRIZ tools can then be applied to find solutions – for which there is also a solution bank.

As can be seen from Figure 12.4, the TRIZ process is quite intricate and has a huge toolkit of analytical methods, which are based on the patent analysis. It clearly requires extensive training and practice to be effective, but it does have a strong and enthusiastic following from its adherents.

12.11   Concluding Points

The aim of this chapter has been to point out the frequency and importance of decision-making and problem-solving in technology development work. Decisions or choices must often be made under pressure with imperfect information and small sample sizes. Even without these pressures, decision-making is subject to a wide range of human foibles, biases, and fallibilities. Framing and structuring a problem or decision in the right way is the first step to improve the quality of the process. At the same time, an awareness of these issues, and some of the common pitfalls, might give pause for thought rather than just blundering on. For everyone involved in work of this type, some study of this topic is likely to be beneficial. An appreciation of statistical methods and thinking is particularly valuable, especially in terms of reacting to 'bad news' and distinguishing whether something has fundamentally changed or whether the bad news is just an extreme version of what is already known. Finally, clear methods of presenting numerical data can often make the difference between acceptance or rejection of potentially sound ideas.

References

This book by W. Edwards Deming is one of the classics of management thinking. It places an emphasis on statistical thinking but, as much as anything, is about management philosophy. Deming still has a strong following through the Deming Institute:

12.1 Edwards Deming, W. (1982). *Out of the Crisis*. Massachusetts Institute of Technology.

More about how the human mind operates is described in:

12.2 Kahneman, D. (2011). *Thinking, Fast and Slow*. New York: Penguin Books.

The same author has also written about the differences between Asian and Western thinking patterns:

12.3 Nisbett, R. (2015). *Mindware, Tools for Smart Thinking*. Farrar, Straus & Giroux.

This book has been referenced in an earlier chapter:

12.4 Nisbett, R.E. (2003). *The Geography of Thought: How Asians and Westerners Think Differently – and Why*. New York: The Free Press.

For a definition of 'statistics', see:

12.5 http://www.merriam-webster.com/dictionary/statistics.

Thanks to technology, it is possible to read not only the full remarks by Lawrence Summers in the twenty-first century but also the entire text of John Snow's great work on cholera from the nineteenth century:

12.6 Summers, L.H. (2012). What You (Really) Need to Know. *The New York Times* (Jan. 20): https://www.nytimes.com/2012/01/22/education/edlife/the-21st-century-education.html.
12.7 Snow, J. (1855). *On the Modes of Communication of Cholera*. London: John Churchill https://archive.org/details/b28985266.

This publication contains some interesting material on how to present data:

12.8 Tufte, E.R. (2001). *The Visual Display of Quantitative Information*. Cheshire, CT: Graphics Press.

A3 problem solving is described in more detail here

12.9 Shook, J. (2008). *Managing to Learn: Using the A3 Management Process Pap/Chart Edition*. Cambridge, MA: Lean Enterprise Institute.

Finally, the TRIZ method is unravelled in:

12.10 Cameron, G. (2010). *Trizics: Teach yourself TRIZ, how to Invent, Innovate and Solve "Impossible" Technical Problems Systematically*. CreateSpace.

13

Improving Product Development Performance

13.1 Introduction

The previous chapters in this book have been concerned with the approaches and methods for developing new technologies or new products. They have laid out the framework within which this economically important activity should be structured and managed. New or emerging organisations can use the preceding material to help them start on their journeys and, hopefully, achieve results more quickly. However, most organisations are not starting with a blank piece of paper. They already have programmes of work in this area and are asking the question, 'How can we improve our product development performance'?

The art of bringing about 'improvement' or 'change' is a well-researched topic. There are many formulae on offer and the many experts on the subject point out that most change initiatives unfortunately fail. Initiatives in the field of product development are no exception to this point and arguably are more difficult to undertake given the long timescales of the activity and the intangible nature of the work.

However, the potential rewards are considerable: an effective technology and product development system is a tremendous competitive advantage that other companies will find difficult to replicate. It is also possible to build up improvements in performance over time, based on a combination of small-scale, local improvements alongside strategic, company-wide programmes. As with most initiatives in business, persistence and determination are the most important factors.

This chapter suggests how organisations might approach this task.

13.2 What Type of Organisation Are We Dealing With?

Organisations working in the technology and product development field vary from the small start-up to the large, well-established corporation. Their

Managing Technology and Product Development Programmes: A Framework for Success,
First Edition. Peter Flinn.

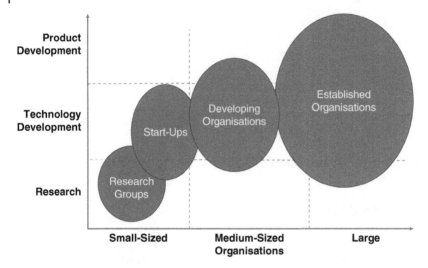

Figure 13.1 Differing forms of organisation.

activities also range across the technology maturity scale. This is illustrated in Figure 13.1.

As can be seen, an organisation might be a:

- Research group, such as in a university
- Start-up or spin-out company
- Established but developing company
- Large, well-established organisations

Each of the different types of organisation can improve its product development performance but the focus of improvement efforts will be different, in line with the role and maturity of that organisation.

The material in this chapter provides guidance on improving business performance generally, but with an emphasis on technology and product development activities. Large organisations will need structure and planning to be effective. Small organisations can use the same principles but can act in a more agile fashion.

13.3 Structuring Improvement and Change Initiatives

There are many ways of managing improvement programmes. The author has direct experience of several, laid out as a series of steps varying in number from about 5 to 14. It is good to have a systematic approach, and the eight-step model proposed by John P. Kotter (Ref. 13.1) of Harvard Business School can be very effective:

1) Establish a sense of urgency.
2) Create the guiding coalition.
3) Develop a vision and strategy.
4) Communicate the change vision.
5) Empower employees for broad-based action.
6) Generate short-term wins.
7) Consolidate gains and produce more change.
8) Anchor new approaches in the culture.

The model is aimed at larger corporations, but its principles can be applied by any size of organisation.

13.4 Diagnosing the Current Situation – Generating Urgency

A good first step is to understand the strengths and weaknesses of the organisation's current technology and product development systems and use this as a start point for the improvement programme. This will provide a focus for the work and therefore help the organisation to bring a sense of purpose, which, in turn, will drive urgency based on potential results. Figure 13.2 provides a framework for understanding the current situation in a company.

The primary relevance of this grid is to established companies, but it does also point out the trajectories that developing companies might inadvertently follow, if they do not make a conscious effort to do otherwise. The idea of this grid came from the Quality Maturity Grid in the well-known book by Philip Crosby, *Quality Is Free*, published as long ago as 1979 (Ref. 13.2). The main usefulness of the grid is in promoting discussion, rather than being a rigid and absolute tool.

Having understood in qualitative terms an organisation's current situation, or where it could potentially be, it is helpful to creating urgency and purpose if some quantitative evidence could also be estimated. This could include:

- Typical product development cycle times compared with competition
- Lost sales or profitability due to slow or late launches
- Additional cash requirements due to longer cycle times
- Missed seasonal launch opportunities

A combination of qualitative and quantitative evidence describing current inefficiencies is a powerful way of communicating the urgent need to improve to all levels of an organisation from its CEO through to its junior staff. It also provides a means of measuring future progress.

Attribute	Under-developed	Developing	Competent
1. Inter-departmental working	Inter-departmental blame and rivalry, 'over-the-wall' mentality, activities undertaken serially	Close but formal working, wider company objectives understood	Close and cooperative working relationship, shared goals, activities undertaken in parallel
2. Definition of market needs	Customer needs defined by marketing based on their views. Little or no direct contact with product development team	Customer clearly identified and market needs agreed across the company initially and then fixed	Customer dialogue built into product development activities. Requirements adapted as work proceeds
3. Development of new technology	No clear view of the maturity of technology planned for new product programmes. Development done on the project with many setbacks	New technologies identified before adoption on a launch project and a broad risk assessment taken	New technologies fully investigated, developed and signed off before adoption on a launch project
4. Product definition	No process for consolidated formal definition of the product	Formally defined and circulated widely	Formally defined, circulated and frequently reviewed and adapted
5. Engineering systems	Product information developed by engineering group but adapted informally by other groups	One common company-wide database of product information with formal change control	One common company-wide database of product information updated rapidly by consent
6. Development and validation by analysis and testing	Limited development before launch. Frequent service problems	Full development programme, functionality proven before launch	Full development, sign-off and reliability trials before launch
7. Manufacturing	Brought into project quite late, design adapted reluctantly for ease of manufacture if essential	Formal manufacturing reviews built into programme from early stage	Close working from outset of project, commonly understood strategy and objectives
8. Supplier management	Contractual relationship	Close dialogue with suppliers within framework of contract	Suppliers built into product development team

Figure 13.2 Product development maturity grid.

13.5   Organising a Way Forward – The Leadership Role

Any significant improvement programme needs to have full management involvement and support if it is to have strategic impact over a long period of time – one of the golden rules of change management. This is especially the case with product development activities, given the cross-functional nature of the process. Change will need the active involvement of engineering, manufacturing, purchasing, and marketing activities as a minimum. The most serious mistake is to see such a programme as the 'engineering department's problem'.

A 'guiding coalition', to use John Kotter's terminology, would have the managerially oriented role of:

- Formulating objectives
- Agreeing priorities
- Assigning people and funding
- Planning the programme
- Reviewing progress

More importantly, it has the leadership role of:

- Creating a vision
- Communicating
- Addressing roadblocks
- Overcoming complacency
- Creating trust
- Celebrating success

The most important attribute of this team is a desire to improve the organisation's performance in the field of technology and product development as a means of enhancing the company's competitiveness. They must then communicate this commitment to their organisation.

13.6   Developing the Strategy and Vision

Any good change initiative or programme is guided by a clear statement of purpose – its vision – laying out what the future should look like, why it would be beneficial to the company, and how it would support customers, employees, and stakeholders. The vision needs to be ambitious but achievable and not couched in overly grandiose or unrealistic terms. Equally, the vision should be neither too vague and unspecific nor so detailed as to limit initiative. A good length is five to six sentences and 100–120 words.

For an initiative directed towards technology and product development work, the vision could include specific reference to topics such as:

- Cycle time to develop new products
- Number of new products entering service
- Reliability of products entering service
- Company spend on research, technology, and product development

By clarifying the direction for change, it should overcome disputes about this direction, although this will only work if the vision is developed as a collaborative exercise rather than being imposed.

It should act as a source of motivation, especially to those who might be disadvantaged by change, at least in the short term. It should also help people to work out for themselves what needs to be done in their area of operation, rather than being expected to be told what to do.

Vision statements are normally associated with broadly based company change initiatives of which improving the product development processes might be part. However, they are just as relevant to something focussed on product development, which would be a wide-ranging initiative anyway, with multiple participants.

Involving many employees in developing the vision promotes a sense of participation and will improve the content of the vision statement. Another helpful way of improving it is to write a longer version, perhaps several pages, through the same process and then go back to the short version for a final edit.

This approach might seem overelaborate for a small organisation with a handful of people. The same process can, however, be easily and quickly followed as a means of clarifying everyone's thinking. It might also be useful in communications with investors or lenders, giving them confidence about the way forward.

It is worth spending some time on this topic, thus doing a thorough job and not being tempted to short-cut the process. In doing so, the words of Gary Hamel (Ref. 13.3), said in the context of strategy, are pertinent:

> Create new passions! In the past, we've driven the emotions out of (strategy). The field is dominated by those who trained as economists and engineers – the two disciplines which have the least grasp of what it means to be a human being.

13.7 Communicating the Vision

A wide-as-possible understanding of the initiative is clearly vital, whatever the size of the organisation. If it is not done properly, people will make their own, usually erroneous, assumptions. Communication is often thought of in terms of briefing meetings, notice boards, intranets, and newsletters. There is a place for these, but they are amongst the weakest means of communications as they lack human content and may be perceived as corporate propaganda. However, some effort is suggested in this area.

It helps, of course, if the message can be expressed in simple, straightforward terms, and there is no harm in frequent repetition.

There are then three ways of communicating the intent more strongly:

1) *Training.* Everyone (absolutely everyone) involved or affected can be given training or awareness about what is planned: its purpose and objectives, the principles on which it is based, the plan of action, the expected results, and the changes required. This is especially powerful if the training is given in person by line management.

2) *Incorporating into everyday discussions.* Rather than being somehow a separate activity, the improvement programme can be built into everyday activities, interactions, and discussions.

3) *Management behaviour.* The real test from an employee's viewpoint is how management behaviour actually changes (not how it says it is going to change). For example, if the company has a history of releasing underdeveloped products onto the market, then postponing a launch until the product is right, at some cost or embarrassment, would be taken as a real signal of intent.

The overall tendency is to underestimate what is needed in the area of communicating the vision, and few managers have the charisma to influence solely through charm and personality. However, leadership by example will overcome much of this.

13.8 Empowering the Organisation

Improving product development performance usually involves speeding up cycle times, placing greater emphasis on risk identification, and seeking a better-optimised product. This, in turn, requires, or brings about, greater devolution of responsibility, and hence more initiative at the working level, with less dependence on supervisory action. It almost certainly will involve greater teamwork among people from different specialisations within an organisation. The term *empowerment* is sometimes used to describe this change – a description that reeks of management-speak but that is nonetheless accurate.

Bringing about such significant changes will need the active involvement of a wide range of people. Quite often, barriers to change appear at this point, either consciously or unconsciously, as people defend the status quo and their own 'territory'.

One of the biggest obstacles of this type is often the area of cross-functional working, which is fundamental to successful product development projects. For example, a wish to select the lowest purchase cost, a typical target for purchasing departments, may conflict with product performance, cycle time, or manufacturing efficiency; having to escalate trade-off decisions some way up an organisation will either slow progress or result in no decision at all.

Another source of problems may be management information systems that emphasise the wrong parameters or fail to measure the right parameters at all. Incentive, appraisal, and personal performance systems may also encourage the wrong approach.

Middle management activities also often have to change, usually from a supervisory, decision-making to a coaching, facilitation, and unblocking role, which may stretch the capabilities and natural skill sets of those involved. These requirements place more emphasis on social, influencing, and behavioural aspects of those involved, rather than largely technical skill sets. Many will require help and coaching of their own to achieve such changes and, for some, it may be too much. These are issues better confronted than left to fester.

The overall point is that, no matter how well the vision is communicated, improving product development performance will require changes to how people operate at the working level, how they are led and supervised, and how they are measured and rewarded. Improvement programmes will therefore require strong and determined leadership if they are to have significant impact.

13.9   Generating Short-Term Wins

Few change initiatives, no matter how well thought out or communicated, can survive for long without some tangible evidence of progress. This is a natural consequence of human scepticism and impatience about business change programmes. It affects both employees, although some will maintain their optimism, and investors, whether they are corporate headquarters in the case of large organisations or private investors in the case of smaller companies.

Fortunately, there are many opportunities for small-scale, short-term projects that can both produce useful results and provide evidence of progress. With a little guidance, staff involved in the improvement initiative will come forward with many useful ideas based on their practical experience of how processes currently run and what problems they experience.

Ideas for such projects include:

- Co-locating a small cross-functional team into a project cell to tackle a difficult issue affecting a live project
- Trying a manufacturing and purchasing review of a developing project at a much earlier stage than normal
- Undertaking a technology readiness level (TRL)/manufacturing readiness level (MRL) assessment of a technology project and then acting on the results

- Doing a quality function deployment (QFD) review on an early-stage project
- Putting together a cross-functional team to improve the product change management process
- Doing a virtual reality servicing assessment of a product at the design stage
- Undertaking a formal and thorough project sign-off before launch
- Reliability testing, with an adequate sample size, of a new product ahead of launch
- Undertaking an early-stage product risk and safety review

Every organisation should be able to find a number of projects of this nature to kickstart the improvement process. The diagnostic activity described above is another source of proposals. Ideally, the projects should involve several areas of the company, including those who do not frequently work together. Such groups will benefit from coaching in problem solving and some initial team building. From this point of view, small projects can be seen as building capability as well as achieving useful results.

Success in the short-term provides evidence of progress and rewards the effort of those involved, thus building momentum and confidence within the organisation. It will help set the direction for future work by identifying areas where the company can achieve results and how they should organise to do so.

It will also be helpful in building confidence from external stakeholders, such as investors, although care must be taken to provide evidence of business performance improvement and not just interesting activity. If tangible results can be shown, these stakeholders will ask for more!

13.10   Longer-Term, Permanent Change

The short-term changes already mentioned will make a useful contribution to the progress of an organisation but will not address root-and-branch improvement in product development performance. A series of small projects will, however, provide good evidence of what further needs to be undertaken and what, more fundamentally, is needed to improve product development competitiveness. They will also provide experience of issues around human aspects of change, especially why people find change difficult and where the roadblocks might be.

From this start point, organisations then need to step up a gear to take on more strategic and fundamental changes. This could be done as a series of projects but over a longer period and with more ambitious objectives. In terms of the projects that could be considered, the following are suggested:

- *Structuring of major projects.* Placing more resource onto early-stage cross-functional activities
- *Organisational model.* Moving away from a functional to a project structure

- *Partnering*. Much closer working between engineering and manufacturing/supply chain partners
- *Facilities*. Investing in improved facilities for analysis and testing
- *Technology programmes*. Building a portfolio of new technologies, developed ahead of specific product programmes
- *Customers*. More early-stage involvement from customers
- *Whole life*. Building up knowledge and experience to understand what is needed for whole-life and connected support of products

Projects of this type provide the mechanics – the systems and processes – for better performance. Human aspects, however, are likely to be just as challenging. As has been noted throughout this book, technology and especially product development are highly collaborative activities. They need input from engineers, marketing people, customers, suppliers, and manufacturing experts, plus others. The easiest way of managing these inputs is as a linear series of activities: the customer specifies what he wants, the marketer consolidates it, the engineer works out a solution, the manufacturer makes it, and so on.

This process gives everyone neat and independent tasks against which they can be measured. Unfortunately, this process does not work. It is too slow to be competitive, and it makes optimisation difficult. If what the customer wants cannot be made cost-effectively, then the cycle starts again, usually amidst a bout of recrimination.

The challenge of bringing about permanent change is the challenge of getting people to work together collaboratively, parallel processing information but within a structured framework. This approach is not something that can be readily imposed; it has to grow within an organisation.

Leadership of fundamental change must therefore provide the conditions for new capabilities to flow but must also let the organisation do the work for itself. As noted above, leadership in this situation is concerned with creating and communicating a vision, overcoming roadblocks and complacency, creating trust, and celebrating success.

13.11   Achieving Permanence

The final question is: how can these changes and improvements be made permanent? The question arises because experience shows that changes of this type can easily regress if the pressure is taken off. Priorities change and people in leadership positions move on, resulting in the emphasis moving elsewhere.

Some of the suggestions made above, such as investing in new facilities and capabilities, will have an obvious permanence and are unlikely to fall into disuse. It is also possible, at least in theory, to write new ways of working into documented procedures – the formal quality systems that organisations of the type described will almost certainly have. However, such procedures cannot readily mandate the constructive and cooperative working, which product development requires. These depend more on the way that people work together than on the processes written down in corporate manuals.

The issue centres on the culture of the organisation. This involves the behavioural norms and shared values of the organisation, which are usually an invisible product of its past. It concerns the pervasive ways of acting in an organisation. New organisations can deliberately create their own cultures, although the result is as much influenced by the background and style of the founders as it is by conscious design. In existing organisations, culture will have been established and reinforced over time. People will have been subconsciously recruited with similar values to those already prevailing, then guided and mentored in those values. Those who are promoted will hold similar values.

All will be well if the existing values support the improved product development approach. Problems will arise if this is not the case. For example, problems will already have been encountered if the prevailing culture was risk averse, political, and slow. Similarly, if managers ran their own independent fiefdoms, then collaboration will already have been found to be difficult. Responsiveness to customers is another critical area.

Achieving permanence for change therefore requires careful attention to the culture of the organisation. In a change programme, it is arguably the last area to change because it depends on first achieving successful results, and it is not something that can be imposed. The key point is that for an improvement or transformation programme to be successful, the changes must be anchored in the organisation's culture. This is especially the case in the area of technology and product development, given the cross-departmental and interpersonal nature of the work.

13.12   Model of Good Practice – Toyota Product Development System

Some leading companies have the reputation of having built up more effective processes than others in developing new products and could therefore be considered as models of good practice from which other companies could learn. Toyota is one example of this category, as measured, for example by its time to market for new products, which was, before its competitors started to learn from Toyota, some 60% faster than the norm.

It is, of course, more well known for its Toyota Production System (TPS), developed during the 1950s, 1960s, and 1970s and made more public from the 1990s onwards. The thinking behind TPS has had a profound influence on manufacturing worldwide.

Their product development system uses similar, waste-avoidance principles but noting, of course, that the 'materials' of product development are information rather than the physical products of the manufacturing world. Their product development system is well documented, not by Toyota itself but by two American authors in the book *The Toyota Product Development System: Integrating People, Process and Technology* by James M. Morgan and Jeffrey K. Liker (Ref. 13.4).

The authors identify 13 principles on which the Toyota Product Development System is built:

1) Establish customer-defined value to distinguish value-added activity from waste.
2) Front-load the product development process while there is maximum opportunity to explore alternate solutions thoroughly.
3) Create a levelled product development process flow.
4) Utilise rigorous standardisation to reduce variation and create flexibility and predictable outcomes.
5) Develop a chief engineer system to integrate development from start to finish.
6) Organise to balance functional expertise and cross-functional integration.
7) Develop towering technical competence in all engineers.
8) Fully integrate suppliers into the product development system.
9) Build in learning and continuous improvement.
10) Build a culture to support excellence and relentless improvement.
11) Adapt technology to fit people and process.
12) Align the organisation through simple, visual communication.
13) Use powerful tools for standardisation and organisational learning.

Most of these points are covered elsewhere in this book, albeit using somewhat different language in many instances. In Appendix 2, there is further expansion of the points above and a cross-reference between each of the 13 points and the relevant section of this book. In context of change management and improving the product development performance of a company, the Toyota system forms a good model, although one that is not easy to copy given that many of the 13 points above relate to the 'softer', cultural aspects of business rather than more tangible or physical activities.

13.13 Models of Good Practice – Agile Software Development

So-called agile methods of software development are another source of ideas and models of good practice, which can be transposed into the world of for engineering. Although the word *agile* can be somewhat overused, and may smack of a management fad, agile software development methods are well documented and address many of the same issues that face engineering product development. Of course, many, if not most, engineering products contain software in some form anyway, so there is a direct link from that point of view.

The origins of the approach can be traced back to the *Manifesto for Agile Software Development* of 2001 (Ref. 13.5), which was a reaction to what, at the time, were seen as overly rigid and micromanaged methods of running software programmes. The approach places emphasis on self-organising and small cross-functional teams, adaptive planning methods, continuous customer collaboration, and delivery of working software in short bursts of activity, or *sprints*.

These principles were built into 12 points that bear quite a striking resemblance to Toyota's:

1) Achieve customer satisfaction by early and continuous delivery of valuable working software.
2) Welcome changing requirements, even in late development.
3) Working software is delivered frequently (weeks rather than months).
4) Close, daily cooperation occurs between business people and developers.
5) Projects are built around motivated individuals, who should be trusted.
6) Face-to-face conversation is the best form of communication (co-location).
7) Working software is the primary measure of progress.
8) Sustainable development is able to maintain a constant pace.
9) There is continuous attention to technical excellence and good design.
10) Simplicity – the art of maximising the amount of work not done – is essential.
11) Best architectures, requirements, and designs emerge from self-organising teams.
12) Regularly, the team reflects on how to become more effective, and adjusts accordingly.

There is some debate about where this approach can or should be applied. It is clearly more suited to relatively small-scale, early-stage activities with open objectives rather than large-scale, closely defined projects in heavily regulated industries such as pharmaceuticals, nuclear, and aerospace. However, many interesting elements of the approach can be used in a wide range of situations.

The terms *scrums* and *sprints* are used extensively in this field, echoing, perhaps unknowingly, a 1986 paper (Ref. 13.6) by Hirotaka Takeuchi and Ikujiro Nonaka in which they drew an analogy between successful product development systems and the game of rugby. The two authors pointed out that rugby games, as with any team sport, operate to only the most general of plans and that results are obtained by passing the ball repeatedly between players as they move forward together down the field of play. As a gross simplification, and with apologies to rugby aficionados, the game alternates between scrums and sprints. Takeuchi and Nonaka found similarities between rugby and the successful product development companies that they had studied in fields such as automotive, copiers, cameras, and personal computers.

Organisations seeking to improve their product development performance can benefit by studying agile software methods. Areas for potential learning could include team structure and dynamics, planning approaches, customer engagement, and a results orientation. Given that most engineering products are now a mixture of software and hardware, the agile approach will cease to be just the preserve of the software industry.

13.14 Concluding Points

As noted above, an effective approach to developing technology and products is a powerful competitive weapon for an engineering organisation. This book has provided an overall framework for such an approach. New and developing organisations can use the material to guide their way forward. Existing organisations may have to change what they already have.

Change within organisations is a well-researched topic but is not easy to achieve. There are structured ways of bringing about change that have been shown to work and to be capable of achieving permanent results. They can, however, seem laborious, with much emphasis on vision, leadership, creating urgency, and communication, rather than the actual improvement activities themselves. This thorough approach has been shown to be necessary to overcome the invisible blockages that change entails.

This is particularly the case with technology and product-creating activities that do not exist in isolation but that involve most areas of companies in this field. A persistent and determined approach, coupled with a good understanding of human psychology, are therefore needed to achieve permanent results.

References

Kotter's book is something of a classic in the field of change management. It is derived mainly from studies of major corporations and their problems of general management. However, the lessons have much wider application.

13.1 Kotter, J.P. (2012). *Leading Change*. Cambridge, MA: Harvard Business Review Press.

Philip Crosby's book is also a classic in the field of quality management.

13.2 Crosby, P.B. (1979). *Quality Is Free, the Art of Making Quality Certain.* New York: McGraw-Hill.

The full text of the interview with Gary Hamel can be found here:

13.3 Jackson, T. (1997). The Management Interview. *The Financial Times* (24 April).

The Toyota product development system is comprehensively described in:

13.4 Morgan, J.M. and Liker, J.K. (2006). *The Toyota Product Development System: Integrating People, Process and Technology.* New York: Productivity Press.

This is a short document that overturned the approach to software development:

13.5 Highsmith, J. Beck, K., and Beedle, M. et al. (2001) Manifesto for Agile Software Development http://agilemanifesto.org

The roots of agile methods go back to:

13.6 Takeuchi, H. and Nonaka, I. (1986). The new product development game. *Harvard Business Review* .

14

Summary, Concluding Points, and Recommendations

14.1 The Rationale for This Book

This book is intended to be complementary to traditional academic engineering education, which teaches engineering theory and supporting practical skills. The book's central theme is how fresh ideas and new technologies can be taken forward and turned into successful products. In this context, it provides a framework and set of principles for approaching this topic. It does not, however, attempt to provide a formulaic approach, which is unlikely to work given that every situation has different needs and therefore requires a tailor-made solution.

In the business community, 'innovation' is widely recognised as a crucial factor in maintaining the competitiveness of enterprises, as well as the spur to creating new businesses. From this point of view, the approach to technology and product development recommended in this book represents the practical means by which innovation can be brought to reality. It is based on the author's own direct experience, mainly in the field of complex engineering products but the principles can be applied more widely. The book should be of interest to engineers in the early stages of their careers, to researchers considering turning their ideas into commercial reality, to managers of non-engineering functions, and to investors. It is deliberately written in nontechnical language.

A key objective of the book is to draw distinctions between 'research' work to identify potential new technologies, 'technology development' to understand fully those new technologies, and 'product development' to deploy new technology in saleable products. The term 'engineering' is used as an umbrella to describe all these activities.

Another key point is that technology is itself bringing about major changes in the art of engineering. Information technologies have radically changed the nature of engineering, particularly how analysis is conducted and how technical information is distributed in organisations. This process will continue as a Fourth Industrial Revolution takes hold. However, the fundamentals of engineering remain the same, such as the need to thoroughly prove

Managing Technology and Product Development Programmes: A Framework for Success,
First Edition. Peter Flinn.
© 2019 John Wiley & Sons Ltd. Published 2019 by John Wiley & Sons Ltd.

products before they go into service – it's just that this can be done more quickly and more thoroughly with new technology.

Most of the individual topics in this book, of which there are many, are already the subject of detailed, single-subject books. The purpose of this book is to tie these subjects together so that the process, as a whole, can be better understood.

14.2   The Engineering Process

The development of technology and products can be considered as a process, albeit one that is not used repetitively, time after time, as would be the case with a manufacturing process. Each project is somewhat different and the timescales in some industries, such as aerospace and defence, can be measured in decades. There are at least three distinct phases to this process, which, as indicated above, can be described as research (which is built on scientific discovery), technology development, and product development.

Taken collectively, work of this nature is a significant activity, typically occupying 1–4% of the GDP of developed nations and totalling close to $2 trillion worldwide. However, its real significance lies in the economic benefit derived from the products that emerge from the process. The economic returns from successful innovation are high. Returns within a company can be in the range 10–30% per annum, and the wider return to society, including spillovers, can be even higher. This is why governments are always keen to promote research and development work.

It is work, however, which is unpredictable, especially in the early stages. When new ideas are being formulated, it is difficult to draw up future timescales and programmes as a series of logical steps, except in the most general sense. This is less the case with the later, and more expensive, stages of product development where classic project management methods can be applied. On large projects at this stage, multiple activities run in parallel with each other, requiring active coordination but recognising that small decisions are constantly required at working level.

There is therefore no magic formula for generating ideas and turning them into successful products. However, broad rules can be followed, and some of the principles of lean thinking, as used in manufacturing, can be usefully applied.

The actual day-to-day process of engineering is essentially an iterative, learning activity. Ideas are formulated, tried, analysed, and tested. The work must also draw on previous experience, requiring an active input from seasoned engineers. Products, or the research and technology development coming before them, can be considered to have a high level of intrinsic risk in their early stages, gradually reducing as more development work is undertaken. The cost of remedying these risks is very low in the early stages, no more than amending a drawing, but is very high if the product is in the prototype stage or, even worse, in service.

The thoroughness of this process determines the eventual reliability of the end product. Public expectation of reliability is far higher than might be appreciated, whether the product is a car, aircraft, or domestic appliance. New firms find it difficult to match the reliability achieved by established companies, derived from their facilities, methods, and experience.

As with any high-level process, rigorous underpinning processes are also needed. The overall framework of ISO9001 provides the starting point. Particular attention needs to be paid to the management of engineering change and to the process for recording and acting upon problems identified during all stages of development and beyond into service with end customers.

14.3   Technology Maturity

The concept of 'technology maturity' is a vital element of the understanding of engineering development. New technology, unfortunately, does not jump out of the box ready to go, moments after having an idea. In fact, the opposite is true: making technology work at the levels of cost and reliability expected by twenty-first century customers is a long process. It is rather like human development, and any attempts at a short-cut are likely to result in a 30-year-old with teenage habits – not a pleasant thought!

To develop this concept further, a mature technology is one that works reliably when in the customer's hands and can be manufactured consistently at the appropriate cost. It is likely to be used in a range of applications by a number of companies. An immature or underdeveloped technology is the opposite of these and will frustrate the end user.

The idea of a technology maturity scale came first from NASA in the 1970s when NASA was trying to understand why certain programmes overran and others did not. It eventually developed a nine-level system, each referred to as technology readiness level, or TRL. This approach has been widely adopted across a range of industries and supports good decision-making: for example, whether to incorporate a new technology on a new product programme, which technology developments to pursue, and whether to invest in a start-up company. Other industries have developed their own TRL scales, each with slightly different characteristics, and the concept has been extended to include manufacturing readiness levels (MRLs).

This then leads to the concept of developing the technology and its manufacture alongside each other. Manufacturing work will always be somewhat behind the technology or product itself but, if the gap becomes too large, there is a danger of an unmanufacturable product emerging. When assessing the maturity of a technology, it is important to look at both the technology itself and the underpinning manufacturing work.

Methods are available for numerical assessment of readiness, and these methods are very useful in providing an objective understanding of the maturity of an idea, and hence what to do next. Quite frequently, certain aspects of new technologies are well developed but other areas are weak. The TRL level assigned to a developing idea is set by the weakest element. For example, the basic performance of an idea may have been thoroughly developed, but its safety performance may have been neglected. Synchronising the various elements of development is an important concept.

In principle, then, advancement of technology readiness is achieved by undertaking increasingly detailed analysis and testing with increasingly representative test material in an increasingly realistic environment. The process for doing this work can certainly be made more efficient, but all phases of maturity have to be worked through. The TRL and MRL scales give a common language for this process and can be used as a means of communication with nonexpert parties such as general managers or investors.

14.4   Aligning Technology with Business Needs

New engineering technology can be developed almost in isolation, but there is then much less chance that it will result in a successful economic outcome. It therefore makes sense, in the early stages of technology development, to think through how that development might be manufactured, sold, and supported, and how it might compete in the marketplace. Just a simple assessment, when a technology is at TRL 2 or TRL 3, will ensure that the development is heading in the right direction. Apart from anything else, this will make future funding more likely whether the development is in a university laboratory or in a company environment.

Manufacturing is a big business worldwide, some 15% of global turnover, so well-presented ideas have plenty of opportunity for development. The idea must, however, be seen in the context of the competition and must have strongly differentiating features and/or must appeal to a very well-defined target market unless it is able to compete on price alone. The latter is possible where the development is taking place in an already strongly cost-competitive business but a new entrant will find it difficult to compete on this basis.

Whatever the competitive stance taken, the technology or product should be capable of articulation through a clear value proposition – as a short summary statement and as a somewhat longer map of benefits against customer needs. This then raises the question of identifying the customer. In some situations, there may be a single, clearly identified purchaser; in others, there may in effect be multiple customers, or the effective purchaser may be an engineer who specifies what will be bought. Identifying the true buyer is essential, and it may not always be obvious who that person is.

If a new technology or product is being developed within an established company, then the route to market will already be in place. Where it is being developed in a start-up situation, there are several ways forward, ranging from developing own manufacture and sale to selling the idea to another business. Understanding the routes forward will, sooner or later, be essential.

Another decision concerns how the idea will be sold: will there be just one product design or will multiple options be needed to satisfy the customer base? Or will each application have to be engineered slightly differently? The ability to satisfy a range of customers is essential.

The product must also be capable of economic manufacture. This means, at a strategic level, employing methods of manufacture that a firm can access, either through its own resources or through its suppliers. At the operational level, it means optimising the details of the product so manufacturing and assembly are easy – difficult-to-manufacture parts invariably have quality problems. This process then extends to service and disposal.

It can be seen from these points that, far from being an isolated activity, technology development needs high levels of collaboration to be effective.

14.5 Planning the Work

Engineering covers a wide range of activities, from small-scale technology development projects to very large product delivery programmes. All forms of projects benefit from at least a basic level of planning – it's a question of how much detail and where the emphasis should lie.

Every project should have six basic documents:

1) Project mandate
2) Project description
3) Milestone plan
4) Project budget
5) Risk analysis
6) Responsibility chart

For a small project, these in total need be no more than half-a-dozen pages; for a very large project, hundreds of pages will be necessary. The process of compiling these documents may be as valuable as the output itself.

Projects can be categorised according to their complexity and level of uncertainty. Simple but uncertain projects, such as early research work, need only the most basic planning, with approximate timescales and a small number of key milestones, but still taking periodic stock of progress, especially as new learning is revealed.

Large, complex projects clearly need very detailed 'classic' project planning and professional management. Before starting such projects, the technological risks need to be brought down to acceptable levels by preliminary technology development work. Otherwise there is a danger of such projects becoming both complex and highly uncertain: a recipe for expensive problems.

There are several ways of organising projects. In small organisations, such as start-ups, the project might in effect be the company. In large organisations, projects can be organised around the company's 'functions' or, alternatively, dedicated project teams can be set up. Both approaches have their advantages and disadvantages.

Projects at their simplest level can be monitored through visual management methods, which can be conducted on a frequent, face-to-face basis. Large projects need more formal methods, including periodic stage-gate reviews.

In summary, new technologies and products are delivered through projects: time-bounded activities with specific objectives. As technology advances in maturity, these projects become more structured, more complex, and more commercially focused, with an expectation by investors that results will be achieved. The basic disciplines of project management are valid at all phases of development: the 'fuzzy front end' through to multimillion-pound commercial projects.

14.6   Creating the Concept

Arguably, the most interesting part of engineering is having ideas and turning them into new product concepts. This phase of work brings together future market needs, new technological possibilities, and economic viability. As with other phases of work, early-stage development is an iterative mix of creating ideas, matching them with gaps in the market and testing whether the solution is likely to work financially.

Some ideas will simply be incremental developments of those that exist already, such as small-scale improvements on last year's product. Much less frequently, radically new ideas emerge and create markets that just don't exist currently.

Ideas can come from a variety of sources: company engineers or salespeople, long-range technology forecasts such as technology roadmaps, other sectors of industry, research engineers in universities, start-up companies, suppliers, or private individuals. Research has shown that three factors tend to determine the success of new products:

1) The superiority of the product in terms of the features it embodies
2) The extent to which customer needs have been investigated in detail
3) The amount of effort invested in early-stage product development

Customer data gathering, in detail, and customer understanding are clearly major factors at this stage, and two models can improve the effectiveness of concept development. The Kano model is often used to guide the understanding of customer needs, and then the quality function deployment (QFD) model is a powerful way of linking customer needs to engineering design detail.

Early-stage technical work forms the foundation of future development: it develops a concept that will appeal to customers. It must also identify critical issues and risks and do sufficient work to show they can be overcome subsequently. A parallel and realistic financial evaluation is a further, important element of concept development.

This is also the stage where intellectual property (IP) protection should be put in place. It could take the form of patents but could take other forms such as copyright and trademarks. Premature disclosure of ideas at this stage should be avoided, as this can prejudice IP protection.

Some level of public funding may be available for early-stage work, especially in the research and concept development phases. There are drawbacks to using these sources but they can provide welcome injections of cash to new organisations.

Concept development suits a small, multifunctional team environment – the work is not easily subdivided and is fast changing. Formal documentation of the work is helpful in as a means of capturing what has been done and as a discipline to ensure that the concept has been fully thought through with no inconsistencies.

14.7   Identifying and Managing Risks

Whilst developing a new concept represents the most interesting part of technology and product development, the enthusiasm for the new must be tempered with a counter-balancing consideration of the risks that something novel might introduce. Such risks can take many forms, from a simple failure to work quite as planned through to outright catastrophic failure leading to loss of life.

Within this context, the risk management approach identifies all possibilities of failure and evaluates them according to their likelihood and to the severity of their consequences. Action is then taken in proportion to these factors. It is through this approach that consumer products such as cars have such high levels of reliability, and rather hazardous activities, such as flying at 500 mph at 35 000 ft, are regarded as everyday occurrences.

The basic root causes of risks and failures are relatively straightforward, and include design-related issues, defects introduced through manufacturing, mechanical failures, electronic component failures, and software design malfunctions. The aim of much engineering development is to minimise the likelihood of these occurring, noting that complete freedom from risk is unattainable.

There are several well-established ways of evaluating risk, such as failure modes and effects analysis (FMEA) and fault tree analysis (FTA), which were first used in the 1950s and 1960s. Industries in the public eye, such as nuclear, aerospace, process, and oil and gas, are very strong in this area, having suffered some serious catastrophes such as Flixborough, Three Mile Island, *Challenger* space shuttle, and Piper Alpha. These industries have built on the basic methods and introduced quantitative approaches that estimate numerically the likelihood of failure and evaluate the consequences, such as injury or damage, also in numerical terms.

These methods are becoming wider in their application as more products and systems become dependent on software and control systems for their safety. Whether it is something as simple as an automatic door opening system in a supermarket or a control system for a nuclear power plant, 'functional safety', as it is called, is an increasingly important topic.

These thoughts must also be tempered by what is practicable and economical. The ALARP concept (as low as reasonably practicable) has been developed to identify which risks are just unacceptable, which can be discounted, and which should be brought down to acceptable levels. Of course, what is considered unacceptable is becoming stricter over time.

Risk identification and management is one of the primary mechanisms for embodying the lessons of the past. Learning from the failures of the past is central to the engineering process.

14.8 Validation

Engineering validation is concerned with the analysis, modelling, and testing activities, which are used to minimise engineering risks and ensure a reliable product. It covers validation of performance, legal compliance, product life, response to extreme conditions, and reliability. Whatever form of validation is used, problems are identified, causes understood, and solutions tested – essentially, a process of learning. All new developments need a thorough and well-planned validation programme to achieve competitive reliability.

Analysis by engineering calculation, based on engineering theory, is the starting point and is readily applied in the early stages of a new technology programmes. Most theory is available as pre-programmed software. More detailed mathematical models are then used to take analysis to a higher level of complexity and detail, examining complete products, systems, processes, and their performance. The success of both analysis and modelling is very dependent on the correlation that can be built up with real life. Modern software makes it very easy to create realistic-looking models and hence the illusion of accuracy. Engineering companies build up this correlation over time, usually in the form of development codes and resulting in very accurate simulation methods. Start-up organisations will have to develop this capability over time and must take care not to be overconfident as this process evolves.

Trialling by physical testing is the ultimate assessment of a new product and clearly the closest to real life. However, the realism of modelling methods has progressed to the point where physical testing is often used more as a means of confirmation than fundamental development. It should be noted that physical prototypes are often the first time that all a complex product's systems come together, and hence their unexpected interactions can be understood. Testing clearly needs access to facilities, instrumentation, and analysis methods. Performance testing is relatively straightforward but life testing needs accelerated methods, which also need to be correlated over time.

In some instances, e.g. where the product is very low volume or a one-off, prototyping may be difficult or impossible. The sold product must then go through a commissioning period which must be carefully managed to achieve customer satisfaction.

Later in development programmes, numerical measurements of reliability can be made if a statistically significant number of products can be made by production methods and operated in realistic conditions – product reliability is set by the thoroughness of the development programme and is not inherent in the design.

14.9 Engineering Delivery

Delivery is concerned with the conversion of a fully developed and validated concept into a formally defined product that can be manufactured with confidence, sold, operated, and retired. The output is information, almost certainly in digital form, such as drawings, bills of material, specifications, and other documents. The engineering function acts as the originator and custodian in most organisations of this data and 'owns' the information, applying formal issue control to it in line with quality management requirements. The data, however, will have been created collaboratively within the organisation and represents an important corporate asset.

In contrast perhaps to the more creative aspects of technology and product development, this is a detailed and exact form of activity, which provides the basis for making known and traceable products.

This activity covers TRLs 7–9 and MRL 5 and upwards. It consumes the majority of the resource and cost of the development programme through detailed design, modelling, prototype manufacture, and test work. Given the amount of resources consumed and the number of activities undertaken, it requires careful planning of the key milestones using classic project management techniques.

There will still be some learning and iteration during this phase of work, but it will be containable if the product specification and technology have been properly researched in earlier phases. However, with inputs from design engineers, manufacturing engineers, suppliers, and other parties, close team-work is needed, backed by responsive but formal change control. Detailed management responsibility should be delegated to team level, where most of the new information originates and where solutions to problems can be found. Co-locating multifunctional teams on either a periodic or permanent basis can have a big effect on the speed of the work at this stage. Good systems in terms of progress tracking, learning points identified and closed, and accessible product databases have a similar effect.

Specialised resources can be troublesome bottlenecks, slowing down this phase of work. These could take the form of specialist engineers, managers for sign-off, analytical resources, or test facilities. Conscious management of bottlenecks is recommended, and it should be appreciated that there is a trade-off between utilisation and throughput time, with high utilisation causing surprisingly long queues of work.

Towards the end of this phase of work, a formal sign-off process should be used to ensure that all requirements have been met, all learning points have been closed, all reviews completed, and legislative requirements met. The prod-uct may then be 'released' without conditions or it may be concluded that a conditional release can be given pending completion of certain tasks.

This is an important decision point at which an organisation commits to vol-ume production and all that it entails.

14.10   Funding the Programme

All developments require money, irrespective of the organisation (start-up or established company) or the stage of development. Persuading an investor or a hard-headed corporate finance man to provide this money is part of the charm of engineering. Some form of business-related plan is a prerequisite to any form of development, with the level of detail increasing with the level of planned expenditure. As a minimum, the plan must explain why customers will buy the product, how it will be made and sold, and what resources are required to make progress.

In established companies, funding will generally be provided from sales-generated cash flow, although new money could also be raised. In a new company, personal money, or that of friends and family, is the start point. After that, angel investors might be persuaded to contribute followed possibly by crowd-funding. Venture capital, banks and public offerings come later. The most difficult period for funding is between demonstration of a concept to the point where a new company is generating some level of sales revenue.

Public funding for technology development, and physical facilities, do exist to cover this period, but it is normally available only for the technology itself and rarely for production facilities, for example.

14.11   Running Teams and Working with Partners

The previous material has concentrated on the structure of engineering programmes from early-stage ideas to series production. It could be inferred that the processes used are essentially mechanistic, and therefore easily put in place. In reality, they depend very heavily on human aspects of collaboration and cooperation. Hence, a basic understanding of human dynamics is an essential element of successfully managing such work. It is easy to underestimate the importance of the topic.

The basic building block for this type of work is the team, rather than the individual. Although there may be a central, core team, the wider team could be somewhat dispersed, especially when suppliers, customers, and partners are taken into account. Team building and team dynamics play an important role. This topic is well researched, and the characteristics of both successful and unsuccessful teams are understood in terms of their composition, development, and dynamics. Organisations will find that paying attention to this topic, and consciously undertaking team development, is well worth the effort.

The extent to which suppliers should be brought into the team will depend firstly on the nature of the contractual relationship (are they simply supplying something out of a catalogue or designing something bespoke?) and secondly the preferences of the lead company for supplier partnership. A hands-off relationship with suppliers does not go well with technical partnership. A further variable is introduced when partners are of different nationalities. The way that different nationalities operate can often lead to unnecessary misunderstanding, another aspect of human dynamics worthy of attention.

The leadership of development teams is a very important role and one which requires a particular mix of skills: proactive out-and-about attitude, meeting people, reacting quickly and being highly communicative, dealing with issues competently and quickly. It is not a role where the person concerned can hide in an office. Other roles require somewhat different mixes of personality traits and there are several models of personality that can be used to match the individual to the role, where there is the flexibility to do so.

It can be seen that some attention needs to be paid to selecting people for development roles and a structured approach to this topic (there is more detail in Chapter 11) improves the likelihood of making good selections. Personal development is also an extremely important topic and is principally gained through project experience, ideally supported by guidance and mentoring. This is an area where individuals must take increasing responsibility, given the relatively short life of organisations and the relative absence of 'jobs for life'.

In summary, human dynamics plays an important part in the running of engineering activities, which are often complex endeavours, undertaken by groups varying in size from a small handful to several thousands. As always, the technology leader has to take a pragmatic view of how to develop and use an understanding of these complex topics.

14.12 Critical Thinking

Decision-making and problem-solving are central to technology development work. Decisions or choices have frequently to be made, under pressure, with imperfect information. These decisions in this field may also have long-term consequences, given that many products last for 10–30 years. Proficiency in 'critical thinking' is therefore an important quality in arriving at good decisions. It requires a focus on: formulating problems, gathering data, weighing alternatives, and reaching sound conclusions with an eye to statistical thinking.

Even without these pressures, decision-making is subject to a wide range of human foibles, biases, and fallibilities. There are about six well-described biases that affect human decision-making; for example, the confirmation bias leads us to note information that supports our point of view but that subconsciously filters out information which does not do so. Awareness of these biases is helpful in reaching the best decision, especially when under pressure. There should also be an awareness that there are two distinct thinking modes – one is fast and is designed to make quick decisions to avoid danger whilst the second is slow and considered. Fast decisions have their place, but they tend to be somewhat imperfect.

There are also clear differences between the mind-sets of different nations, with the West being driven by explicit logic and the East being more open to ambiguity. A combination of both approaches is the ideal in this field of work. At the same time, when working with different nationalities, it must be appreciated that significant cultural differences exist between different countries, companies, and individuals.

Framing and structuring a problem or decision in the right way is the first step to improve the quality of the process. There are some simple methods, such as the A3 structure described in Chapter 12, which can be used to pursue problems to logical conclusions and to involve a wider team in the process to broaden the experience brought to bear on the problem. There are also methods of creative problem-solving, such as TRIZ, which can be used to guide more creative work, usually in the earlier stages of technology or product development.

Statistically based methods have had a big impact on manufacturing effectiveness, improving quality and consistency, and reducing cost. There are areas of product development where the same is possible; for example, in analysing customer data-sets or reliability or warranty information. However, sample sizes during, for example, laboratory or prototype testing are likely to be quite small. There are however methods of analysing trends even with small sample sizes and, more importantly, applying the principles of statistical thinking.

For everyone involved in work of this type, some study of this topic is likely to be beneficial. As always, there is a trade-off between quality and speed. However, competence and structure in decision making is likely to yield better decisions in less time.

14.13 Improving Product Development Performance

Having laid out in the earlier parts of this book how the processes for developing new technologies and new products should be approached, the question then arises as to how companies should set up effective systems or should tackle the improvement of their current systems.

For early-stage companies – start-ups or spin-outs, for example – there will be no well-established product development system in place and the concentration of effort will be on developing new technologies and their applications. They should do this in a way that lays the right foundations for future work. This can be done by paying particular attention to such areas as formally recording all engineering data, results, and reports, gaining an early understanding of manufacturing and supply implications, understanding the market, costs, and business drivers for the new technology, putting in place systems for recording learning points, and planning the structure of future phases of work, especially timescales and costs.

Established companies wishing to improve their product development systems should consider a well-structured improvement process. Several models for this exist and an eight-step process is described in the main body of the book. An important point to bear in mind is that developing new technologies and products is a company-wide activity that has a huge bearing on the future progress of an organisation. It is not confined to one department – 'engineering'. With wide involvement across an organisation, long-term success will only be achieved with clear and determined vision and leadership. This needs to be accompanied by effective communication and some level of training for all involved.

The improvement process itself can include a number of short-term projects, which should produce early results, combined with long-term strategic projects. A good start point is a diagnostic survey of the existing system; a potential structure for this is described.

Achieving permanent results will depend on there being a receptive and supportive culture within the organisation, or moving the culture so that this is the case. Otherwise, progress can easily be lost.

However, the potential gains are well worth the effort. An effective system for developing technology and products is a tremendous competitive weapon that other companies will struggle to copy.

15

Future Direction

15.1   Introduction

So, what does the future hold for engineering development?

The means by which new, technology-based products are created, and then brought to market, have a history extending back at least 200 years, or much longer, if taking a broader view of what constitutes an engineering 'product'. From one point of view, this creative process has remained unchanged over this period. It comprises: find out what might be useful, create an idea, refine and test it, make it, then monitor it in use. However, nothing in life is simple; technology has advanced in leaps and bounds, customers have become more affluent and demanding, and the world has become a smaller place, creating bigger markets but also facilitating competition.

The process of developing new technologies, and hence new products, is likely in the future to be influenced by a continuation of these trends, including, more specifically:

- *Advances in the methods used in the product development process itself*. This includes how new technologies and products are defined, analysed, modelled, developed and tested before entering service.
- *Adoption of new materials and product technologies.* Continuing the trend of decades, if not centuries, involves potentially new combinations, such as conventional engineering, biotechnology, and intelligent systems.
- *Constraints and opportunities.* Both come in relation to energy, environmental factors, and materials availability.
- *Developments in manufacturing systems.* These are becoming more complex and distributed but are capable of producing a much wider variety of products, in turn, affecting product development processes.
- *Increasing demands from customers.* Demands could come from individual consumers or businesses, greater competition, and new models of business.
- *Product connectivity.* Products are connected to either the owner or the manufacturer from the field of operation.

Managing Technology and Product Development Programmes: A Framework for Success,
First Edition. Peter Flinn.
© 2019 John Wiley & Sons Ltd. Published 2019 by John Wiley & Sons Ltd.

These overlapping drivers of change are discussed briefly and individually in this chapter. Their likely impacts on engineering development activities are suggested.

15.2   Product Development Technologies

Product definition information has completely migrated from paper-based to digital forms, and integrated into business-wide systems. However, there is more to come from digital methods of product definition. They will have to be developed further to cope with more complex and distributed methods of design and manufacture combined with fluid supply chain partnerships. Methods of workflow management and data sharing will support the development of global teams.

The products themselves will exhibit more variety to cater for individual customer needs, including the provision of bespoke designs where the manufacturer provides a design and visualisation sales tool, much as is currently done with domestic kitchens, rather than an off-the-shelf solution.

The creative side of product design will be enhanced and accelerated by further developments and cost reduction in visualisation tools, including augmented or virtual reality technologies. The same technologies could also be used to optimise design for manufacture and as a training tool for manufacture and service.

There is also the possibility of wider use of automated design optimisation tools where the engineer specifies the envelope for a component design and the design system runs through multiple iterations to find the optimal solution. Known as 'generative design', this process can mimic biological evolution but over a period of seconds rather than millennia. This would improve upon design optimisation by manual trial-and-error. Whether it could be extended into the use of artificial intelligence, including the ability to learn, remains to be seen.

Modelling and simulation technologies are at the heart of product development and have progressed from relatively simple applications, such as static structures, to more complex problems, such as fluid flow and system dynamics. These technologies will continue to develop, enabling more complex analyses and the ability to explore the interaction between separate systems – previously the domain of the first physical prototypes. 'Multi-physics' methods are developing where, for example, structures, fluid flow, and thermal analyses can be conducted in parallel using the same basic models. The aim is more thorough analysis ahead of physical tests and analysis of difficult or dangerous conditions. As always, a challenge will be identifying the correct way to construct accurate models and verifying the accuracy of simulation, not just producing results which present well.

Further developments aim to make complex analyses more usable by nonspecialist staff (analysis specialists are one of the most common bottlenecks in product development projects), which would also encourage use in earlier stages of projects and use more frequently, which in turn would help learning. This is sometimes described as 'democratising' modelling and simulation.

Products in service will generate much more operational data, which clearly is useful in optimising development processes – see Section 15.5.

Much is made of the development of autonomous products. As well as solving the technical problems of how products might have the intelligence to 'self-drive', ways will have to be found to validate and gain regulatory acceptance of products with such potentially unpredictable characteristics and new modes of failure.

Further challenges will come in the area of human organisation: how to integrate more complex and dispersed participants in a development programme; how to develop the skills for a wider range of technology integration, requiring in-depth expertise on the one hand and broader system integration skills on the other; and simply how the waste-avoidance principles of 'lean thinking' can be applied to these complex human systems so that they perform efficiently and are capable of being managed. It will be interesting to see whether simulation and modelling can be used to model engineering programmes in much the same way as factory production systems can already be modelled.

15.3   New Materials and Product Technologies

New technologies will continue to be generated and assimilated into products, as has always been the case in the past. Whilst this process will present challenges on a case-by-case basis, the practical adoption of new technologies is not fundamentally new and is one of the basic challenges of engineering. The area where some difficulty might be expected is where technologies are combined in unusual ways. For example, grafting intelligence or autonomy onto otherwise predictable mechanical or electrical products cuts across conventional thinking and methods of analysis. Some medical products, an expanding market, might include a biological element, 'grown' rather than manufactured. Similarly, algae-based solar cells and fuel cells have been proposed, also combining biology with electrical engineering. Nanotechnologies represent another radically different area that will be combined with conventional technologies.

15.4   Energy, Environmental, and Materials Availability

There will be continuing pressure to reduce the energy consumption of products during their service, driven first by operating economics and second by legislation to encourage low-carbon or low-energy solutions

Environmental legislation will continue to tighten, placing lower limits on emissions such as carbon dioxide and oxides of nitrogen as well as reduced limits on hazardous wastes from manufacturing processes and limitations on the use of potentially carcinogenic materials in the products themselves. New hazards may be identified and may become the subject of future legislation.

Relatively scarce or expensive materials are now being more widely used in applications such as electronics, photovoltaics, catalysts, magnets, motors, generators, batteries, fuel cells, and mobile phones. The materials used include rare earth elements, rhodium, indium, gold, palladium, and platinum. The quantities per unit are very small, but their role is critical to the functioning of the devices. It is unlikely that supplies of any of these materials will actually run out. However, prices could rise unacceptably, some are by-products of extraction of more common materials, and some have security of supply issues, coming from politically difficult countries. The latter point also affects some less-scarce materials such as lithium.

These factors will drive programmes of material substitution. They will also encourage closed-loop approaches where these materials are recycled with minimal loss, rather than being 'single use'. In fact, there will be growing business and engineering opportunities in developing systems for recovering these scarce materials such as already exists for recovering platinum-group metals from automotive catalysts.

15.5   Manufacturing Systems

The future development path of manufacturing systems over the coming decades is a huge and complex subject in its own right, and the following paragraphs can barely scratch the surface of it. At the level of the machine or cell, information processing is assuming as much importance as physical, manufacturing processing; the term 'cyber-physical systems' is used to describe facilities exhibiting this combination. Automation in this context should be considered as a complete system, rather than the individual machine or cell. Adaptive methods of manufacture, including but not limited additive/3D-printing, will replace 'hard' automation.

The driver for this is the further subdivision of supply chains to provide distributed, adaptive, and multipurpose manufacturing rather than the monolithic approach taken traditionally to high-volume/low-cost production. From a customer's viewpoint, it will provide the opportunity for tailor-made products at mass manufacturing prices.

These developments will clearly have their impact on how the products, which are processed through the manufacturing systems, are designed in the first place and how the product information is presented. Whilst design-for-manufacture, in the sense of optimising designs to the physical manufacturing process, will still be important, products will also have to be configured in a way that optimises information flows – 'design for IT'. Designs and information will have to facilitate local manufacture, optimise material and energy use, and support multiple suppliers. More particularly, rather than there being one design to be made by the million, ways will need to be found of designing (and developing, proving, and signing off) a wide range of similar products, capable of personalisation.

A McKinsey report of 2013 (Ref. 15.1) summarises this very well:

> *The new era of manufacturing will be marked by highly agile, networked enterprises that use information and analytics as skilfully as they employ talent and machinery to deliver products and services to diverse global markets.*

15.6 Customer Demands

At the simplest level, customers will continue to expect increasing levels of performance and feature levels at reducing levels of real cost, whilst engineers will continue to develop technologies that can meet these needs, or create new ones. As noted above, customer choice and bespoke designs are further areas of increasing expectation. However, another important driver of change will be new models of ownership.

Industrial customers are increasingly buying the product as a service where the supplier provides the effect of the product, rather than the product itself. Examples where this is currently practised include railway trains, compressed air, document copying, lighting, and aero-engines. The customer buys, for example, compressed air by the cubic metre, rather than investing in a compressed air system, which becomes the responsibility of the supplier.

In technology development terms, the product must then be instrumented and connected back to the supplier to provide usage data and be developed alongside the appropriate business model and financing arrangements. In some industrial sectors, service revenue now makes up 50% of total revenue. Changes in business models will drive product development organisations into wider involvement in business matters, beyond developing the product itself.

Another example is shared ownership models where the product is hired – already common practice in many sectors, such as automobiles, but extending into areas such as bicycles. In this situation, the product is likely to be used more intensively and possibly be subject to more misuse, requiring the design rules for the product, and its development programme, to be amended to meet the new usage situation.

15.7 Connected Products

To an increasing extent, products will be connected to manufacturers, operators, or owners via the Internet using computing devices embedded in those products, thus enabling them to send and receive data. The 'Internet of Things' will enable a wide range of new possibilities, including health monitoring, failure prediction, performance optimisation, and remote management. These capabilities must clearly be designed into products from the outset and the data communication systems developed alongside the product itself.

These technologies will obviously generate large quantities of data – 'Big Data' – where the volumes of information exceed the capabilities of traditional methods of analysis. There will therefore be an increasing role for 'data scientists' in product development organisations.

If ways can be found of structuring and analysing this information, it could provide the basis of better-optimised designs based on a more thorough understanding of usage. However, care must be taken before design rules, built up over many years of practical experience, are rewritten. Regulators, in particular, will need to be persuaded that new approaches are safe. At the same time, these technologies will present cybersecurity challenges (see, e.g. Ref. 15.2).

15.8 Concluding Points

New technologies will continue to drive the creation of new products and services, as has been the case for decades, if not centuries. They will also impact the processes by which those products and services are created. As well as affecting methods of definition, analysis and testing – the bread and butter of product development – they will also affect, rather fundamentally, the way that products are made, their subsequent connectivity, and the services associated with them. This, in turn, will broaden the process of product development from the simply designing and making a product towards the creation of partnerships over the lifetime of the product.

Hence, the skills of product development will become correspondingly broader. At the same time, more specialists will be needed in new areas such as data science. In an ideal situation, this will need what are sometimes described as 'T-shaped' engineers – capable of handling a broad range of disciplines but competent, in-depth, in one speciality. Whether these two needs can ever be met in one person is a matter of debate, but part of the fun of engineering is grappling with incompatible requirements!

References

There is relatively little material about how product development processes are likely to evolve over time, but there is a lot about future manufacturing systems, which can be used to judge the likely path for manufacturing businesses generally. Two which are worth reading are:

15.1 Manyika, J., Sinclair, J., Dobbs, R. et al. (2013). *Manufacturing the Future: The Next Era of Global Growth and Innovation.* McKinsey Global Institute Available at: http://www.mckinsey.com/insights/manufacturing/the_future_of_manufacturing.

15.2 Foresight (2013). The Future of Manufacturing: A new era of opportunity and challenge for the UK – Summary Report The Government Office for Science, London Available at: https://assets.publishing.service.gov.uk/government/uploads/system/uploads/attachment_data/file/255923/13-810-future-manufacturing-summary-report.pdf

In addition, some of the engineering software and analysis companies have 'white papers' on how they see their products evolving over time.

Final Thoughts

The work of developing engineering technologies and products can be a fascinating and creative pursuit. Anyone who has contributed to a successful product or installation will come across reminders of it over a significant period of time during its everyday operation. If he or she is lucky, the same product may then become a collector's piece and enjoy further, almost never-ending, popularity. The trials of bringing it to fruition will then be long forgotten.

At that point, it is worth reflecting on what lay behind that success. Thoughts will turn to particular individuals who had the vision and tenacity to make things happen. Hopefully, what they actually did will have been some of the suggestions made in this book.

The author is always open to further ideas that could be used to improve the content of this book and would be happy to enter into correspondence on such points.

Peter Flinn
Ashbourne, Derbyshire,
United Kingdom
2018

Managing Technology and Product Development Programmes: A Framework for Success,
First Edition. Peter Flinn.
© 2019 John Wiley & Sons Ltd. Published 2019 by John Wiley & Sons Ltd.

Appendix A

TRL and MRL Definitions

(Taken with thanks from: Measuring Technology Readiness for Investment – The Manufacturing Technology Centre & Heriot-Watt University, March 2017 – Ref. 3.8)

The purpose of the attachment is to provide a plain-language description of the technology readiness level (TRL) and manufacturing readiness level (MRL) scales. The starting point for the scales was information published by NASA, US Department of Defense, US Department of Energy, European Commission, and the UK Automotive Council.

A.1   Technology Readiness Levels

TRL 1. This is the first stage of development for a new technology. At this stage, a need for the technology will have been identified, perhaps in the form of a problem to be solved or a market need to be addressed, and the principles of a solution will have been developed. The solution will be described briefly, probably in narrative form or as a feasibility study, and its performance and characteristics will be predicted. This will be supported by sketches, diagrams, and some calculations or simple mathematical modelling. A literature search may identify other, similar applications that provide confidence that the proposal is a feasible proposition. The development work will probably have been carried out in a research, laboratory, or workshop environment. There may also be some initial ideas about how the solution could be manufactured, but no more.

TRL 2. The concept outlined at TRL 1 will be developed in greater detail and more thorough performance predictions will be made. Initial drawings or CAD models of the solution will be produced. Simulations or analysis of critical elements of the solution will also be produced, and physical tests of these elements may be undertaken to correlate the analysis. These physical tests are likely to have been limited to components of the solution rather than the solution as a whole. Areas will be identified to which formal IP protection could be applied. Basic manufacturing feasibility studies will also be undertaken, outlining the approach that might be taken and showing its basic feasibility.

TRL 3. The basic concept of the new technology is shown to be viable ('proof of concept'). On the basis of this, further investment in the next, and more expensive, stages of development might be expected by either the organisation doing the development or by an external investor. The design of the solution will be complete, as an overall system but not the details. This will probably be in the form of a CAD model plus coding of any embedded software. Substantial simulation and modelling will have been undertaken to prove the performance characteristics of the system. This will be supported by practical test work, in a laboratory environment, of a complete system, albeit using relatively crude representations of the final items. This demonstration system is unlikely to work reliably, and could be quite temperamental, but it will show that the concept can be made to work and can achieve something approaching the required level of performance. The development work will also be supported by an initial quality function deployment (QFD)-type analysis, linking the product in detail to the market need which has been identified. Failure modes and effects (FMEA) or similar methodologies will also be started to identify areas critical to the reliability of the solution. Formal IP protection will be sought. Finally, a manufacturing approach will be defined and documented, suited to the volumes of the market identified.

TRL 4. The basic functioning of the proposed new technology will have been demonstrated at TRL 3 but in a laboratory environment and with limited repeatability. The aim of TRL 4 is to build up an understanding of how the technology works so it can be developed and used with much more confidence. The emphasis, therefore, is on practical testing of the complete solution, rather than elements of it, but in a laboratory environment and using components that are not wholly representative of the final solution. Modelling and simulation work will proceed in parallel with the practical work, and this will feed off the practical work. Initial QFD and FMEA studies will be completed and the results fed into the technology design and into the test program. The QFD, or similar, study will require the market application to be defined in some detail. More detail will also be added to the design of the technology, which will take the form of CAD models from which detailed drawings and bills of material could be extracted. Manufacturing work will have identified the basis of a manufacturing process or system for the technology. Formal IP protection will now be in place. There will be a broad indication of the timescale and costs to reach TRL 9.

TRL 5. The work undertaken at TRL 4 should give sufficient confidence for the technology to be taken out of a laboratory or workshop environment for the first time. This could be achieved by installing it on a test vehicle if, for example, it is an automotive, aerospace, or rail technology. Alternatively, it could be installed, on a one-off basis, in something close to the true operating environment but under close control and operated by the technology developer, not the customer. Further details will be added to the design that will continue to be optimised, and one output of this TRL will be a set of design rules that provide the guidance for designing a successful product in the future. At this stage, manufacturing studies should be undertaken in some depth. The manufacturing process routings should have been identified, including any new technologies or hazardous processes. Sources of materials, supply chains, and capacity to produce, including any special materials, should be known. This work should be fed back into the design through formal Design for Manufacture & Assembly (DFMA) assessments, which should optimise the design to achieve cost targets. Any special skill needs for production should be known and the broad programme through to TRL 9 should be confirmed.

TRL 6. At the TRL 6 stage, the technology is ready to be taken out into the field for trials by known end user(s) in a form close to the final end product. The performance of the technology will be measured closely for comparison with the expected performance and benefits. Based on successful completion of this stage, a decision should be made to launch the technology on a specific product platform, or as a product in its own right, by a certain date and within known cost boundaries (product cost and investment). As well as operating the technology under 'normal' conditions, it could also be trialled at extremes, such as overload, abuse, harsh environments, or with unfamiliar operators. Simulation models will continue to be used in parallel with the practical work, in particular to investigate any problems identified. The design of the technology will be stable and detailed at this stage but will be open to detailed modification based on the field trials. Manufacturing planning will have started in detail, and this could impact on the design in terms of DFMA studies, supply source capabilities, and capacities and lead times. Some redesign might be expected to deal with issues related to manufacturing, and these will have to be dealt with in parallel with the field trials and its outcome.

TRL 7. This stage has the aim of confirming that the product embodying the new technology meets all the performance, durability, and reliability criteria specified for that product. It may involve multiple prototypes of either the complete product, and/or elements of it. These prototypes will be as close as possible to the final product but may be produced by low-volume methods or without production tooling. The prototypes will be subject to a wide range of tests that could include coverage of extremes of the operating envelope, overload, abuse, and extreme environmental testing. Issues identified by earlier FMEA work will be signed off as resolved at this stage. The product itself will be fully defined in terms of drawings, bills of material, and specifications, which will be the subject of formal change control to manage the modifications that will inevitably arise from the test programme. Models and simulations from earlier TRLs will be maintained and used in problem solving. Detailed manufacturing planning will run parallel to this activity, and production of prototypes will provide useful learning to guide the full production process. This learning will be captured in formal DFMA processes, which will have been started at much earlier TRLs.

TRL 8. TRL 8 is linked to the establishment of a full production system for the product that will embody the new technology. The product itself will be stable in the sense that the design will not be subject to major modifications as a result of the test programme, which is largely completed at TRL 7. There is, however, the possibility that long-term durability testing, a TRL 7 issue that will overlap with TRL 8, will throw up some issues and similarly early pre-production products may be introduced into the field for reliability trials. There will also be scope for changes and optimisation arising from manufacturing and supply chain development. Full regulatory approval, if needed, will be completed at this stage, which may require approval of the production system – the physical assets, the processes, and the skill base – as well as the product itself.

TRL 9. At TRL 9, the technology that started its journey at TRL1 will be operating successfully in the market with customers and be providing the benefits and capabilities that were originally envisaged. Manufacture will be operating smoothly within acceptable cost, quality, and delivery boundaries. Skills and training to achieve this will be in place. There will be programmes of cost reduction and productivity improvement. The technology will be operating reliably and will be the subject of continuous improvement rather than rectification. Reliability performance, and other measures such as warranty costs and customer complaint levels, will be known and acceptable. The emphasis will be on expanding market coverage and developing new applications for the technology.

A.2 Manufacturing Readiness Levels

MRL 1. This is the first stage of development of the manufacturing processes, which will be used to produce, for sale to the marketplace, a new product based on new technology. It can take place when the product concerned has reached a TRL 2 or TRL 3 stage of its development. At MRL 1, there will be paper-based research of the possible manufacturing methods that could be used, appropriate to the sales volumes envisaged. There will be a broad assessment of manufacturing costs, investment, sources of supply, and timescales. This will also highlight any areas of potential difficulty, e.g. unusual materials or processes, where early action might be needed to forestall potential risks.

MRL 2. More detail will be added, at this stage, to the work undertaken at MRL 1. In particular, initial conclusions will have been reached concerning difficult or unusual materials or processes. Initial drawings and bills of material will be available, which will enable DFMA studies to commence. The product will be at a TRL 3 or TRL 4 stage of maturity so there will be hardware to evaluate, covering both its design and its early-stage manufacture and sourcing. From this work, a manufacturing development programme can be defined. These points will be documented in a manufacturing feasibility and development plan.

MRL 3. At this stage of manufacturing development, the new technology or product will have advanced to the stage where a trial system might have been taken out of the laboratory or workshop environment and be operating in a market environment. There will therefore be a lot of design detail and experience from the manufacture of early components or systems. At this stage, then, manufacturing studies should be undertaken in some depth. The manufacturing process routings should have been identified, including any new technologies or hazardous processes. An initial make-versus-buy study will have been undertaken. Sources of materials, supply chains, and capacity to produce, including any special materials, should be known in principle. If substantially new manufacturing methods or technologies are required, workshop trials will have been conducted to confirm that the planned manufacturing concepts will actually function. This work should be fed back into the design through initial DfM assessments, which should optimise the design to achieve manufacturability. Skills needs for production should be known and the programme through to full production should be confirmed or modified.

MRL 4. An early prototype is likely to have been produced when the manufacturing readiness has reached MRL 4. Hence, the product will be capable of manufacture in a workshop or low volume environment. Problem areas or risks will be known, and the design will have been optimised as a concept for prototype production. This then provides the basis for planning high-volume production, and processes will have started for this, including formal in-house and supplier quality-planning systems. The potential supply chain will have been surveyed, sources of supply and lead times will be known, and make-versus-buy decisions confirmed. Simulation models will have been established for critical, individual processes and for the overall production layout. Workshop trials for any new manufacturing technologies will have proceeded to the point of reliable operation in a high-skill environment. The programme through to full production will be known in detail, including revenue and capital costs for subsequent MRL stages. A high-cost commitment to enter a full prototype development and production planning programme will be made by the company at the end of this stage.

MRL 5. Manufacturing readiness will have advanced to the level where multiple prototypes can be produced by low-volume methods to permit a detailed performance and durability programme to be undertaken on the new product. By committing to build multiple prototypes, supplier company choices for production will in many cases have been made in effect. Part-by-part DFMA studies will be completed during this phase to capture the learning from prototype build. Manufacturing process FMEAs or similar process risk studies will also be completed during this phase. Trials of new manufacturing technologies will be running with production-level operators. The product itself will be fully defined in terms of drawings, bills of material, and specifications, which will be the subject of formal change control to manage the modifications that will inevitably arise from the test and manufacturing review programme.

MRL 6. Having started to produce multiple prototypes and learnt from this experience, the design will now be capable of being produced in volume and most production manufacturing processes will have been defined. New manufacturing technologies will now be capable of production operation, and statistical methods will demonstrate this. Supplier agreements will be in place and long-lead production items will have been identified. Cost analyses will be updated from prototype experience, and overall cost targets will be achievable with actions in place for problem areas. The investment budget will be known on a detailed basis, and a commitment will be made to go ahead with that investment.

MRL 7. Facility design procurement will be undertaken covering capital items, tooling, fixtures, inspection equipment and material handling. Process FMEAs or other risk analysis methods will be completed ahead of this. Simulation modelling will have been completed and critical areas identified. Supplier quality assessment will be complete. Material will be ordered, starting with long-lead items, for facility commissioning and parts schedules issued accordingly. Cost models will be updated as orders are placed.

MRL 8. At this stage, production facilities will be in place and will be commissioned to produce the full range of products in the required quantity. Simulation models will be run in parallel with physical commissioning and used to support problem solving. Parts and products will be inspected in detail to confirm that all requirements have been met. This will apply to both in-house and externally sourced items. Formal certification of processes by external agencies may be required at this stage. Sufficient operators for MRL 9 volumes, and support staff, will be fully trained and signed off as competent.

MRL 9. All production facilities will be fully operational at this stage and have a proven capability to run at the full, planned volumes. Volumes will be gradually built up. Further staff will be recruited to facilitate the expanding output. Process capabilities and facility performance will be closely monitored during the period of volume increase, as will the performance of the supply chain. Costs will be fully known and within target levels.

MRL 10. At MRL 10, manufacturing will be running consistently at the planned volume levels and parameters such as on-time delivery, quality, and cost will be stable within the target parameters. There will be a full complement of trained staff. Statistically, processes will be in control and the emphasis will be on continuous improvement, cost reduction, and productivity enhancement. All facilities will be fully operational and proven.

Appendix B

Toyota Product Development System 13 Principles and Their Cross-Referencing

*(Derived from *The Toyota Product Development System: Integrating People, Process and Technology* – James M. Morgan and Jeffrey K. Liker)*

No.	Principle	Description (from Morgan and Liker)	References in main body of book
Process			
1	Establish customer-defined value to distinguish value-added activity from waste.	Waste takes the form of untimely or incorrect engineering information.	Ch. 4 – value proposition, satisfying a range of customer needs Ch. 6 – identifying customer needs Ch. 8 – validate against customer requirements Ch. 9 – product sign-off
2	Front-load the product development process when there is maximum opportunity to explore alternate solutions thoroughly.	Cross-functional collaboration is needed early in the design process to make sure that later rework, due to poor decision-making in the early stages, is prevented.	Ch. 2 – cost of problem resolution, early problem detection Ch. 3 – technology maturity concepts
3	Create a levelled product development process flow.	A value-stream map should connect important milestones to decisions, information flow and critical meetings – key integrating events.	Ch. 5 – goal-directed management, value-stream mapping, managing, and monitoring projects Ch. 9 – planning and decision making, flow of information Ch. 11 – working collaboratively

Managing Technology and Product Development Programmes: A Framework for Success,
First Edition. Peter Flinn.

No.	Principle	Description (from Morgan and Liker)	References in main body of book
4	Utilise rigorous standardisation to reduce variation, and create flexibility and predictable outcomes.	This can cover a wide range of topics, including design methods, calculation and modelling methodologies, test codes, parts standardisation, and such parameters that are intended to create predictability of results.	Ch. 7 – identifying and managing risks Ch. 8 – standard validation methods Ch. 9 – engineering delivery

People

No.	Principle	Description (from Morgan and Liker)	References in main body of book
5	Develop a chief engineer system to integrate development from start to finish.	The chief engineer is an important role for every product development programme. This is both a project leader and senior engineer in one combined role. He/she represents the voice of the customer and is responsible for the development value stream, from concept to production.	Ch. 2 – governance of the process Ch. 5 – organising for projects Ch. 11 – leadership, selecting people
6	Organise to balance functional expertise and cross-functional integration.	This is the challenge of integrating the chief engineer role into the organisational structure. The chief engineer is responsible for the delivery of the product and the voice of the customer. The functional manager is responsible for the development of his team members.	Ch. 4 – linking engineering to the broader business Ch. 5 – organising for projects Ch. 9 – flow of information Ch. 11 – working collaboratively and team development
7	Develop towering technical competence in all engineers.	Toyota prefers specialists over generalists. Every engineer has to have a clear development path where he learns the specific skills needed to join a development team in a certain role, and the development paths should be standardised.	Ch. 2 – learning cycle Ch. 7 – identifying and managing risks Ch. 8 – validation using experienced, practical engineers Ch. 11 – selecting and developing people Ch. 12 – critical thinking
8	Fully integrate suppliers into the product development system.	Companies should manage their suppliers the same way as they manage their own production. Suppliers' expertise can be very valuable to the design process, so suppliers should be involved from the earliest stages.	Ch. 4 – industry structure Ch. 9 – examples of good & bad practice Ch. 11 –working with partners

No.	Principle	Description (from Morgan and Liker)	References in main body of book
9	Build in learning and continuous improvement.	One of the most important aspects of the project is the reflection afterwards on personal, team and project level. What did we learn? How can we improve our processes? Toyota plans three 2-hour sessions of reflection after each project. This important part of the process is often neglected, although the ability to learn faster than competitors could be the only sustainable competitive advantage.	Ch. 2 – learning cycle Ch. 7 – identifying and managing risks Ch. 8 – validation methods based on earlier experience
10	Build a culture to support excellence and relentless improvement.	A culture is defined by the current generation of leaders and defines which leaders will emerge next. Leaders should therefore set the example of learning and always ask about the improvements.	Ch. 4 – linking engineering to the broader business Ch. 11 – working collaboratively

Tools and Technology

No.	Principle	Description (from Morgan and Liker)	References in main body of book
11	Adapt technology to fit your people and process.	Traditional firms act the other way around and alter their processes to fit a certain tool. A tool or technology should be integrated into a process, and should focus on specific solutions and enhancing people, not replacing them.	Chs. 1 and 2 – impact of technology Ch. 9 – engineering delivery Ch. 13 – improving product development performance
12	Align your organisation through simple, visual communication.	There are multiple tools that can support communication between team members and between teams: Hoshin Kanri for goal alignment between departments, team boards and for development teams; Obeya for alignment between team members; A3 methods for problem solving on individual level.	Ch. 5 – managing and monitoring projects, project mandate Ch. 12 – problem-solving
13	Use powerful tools for standardisation and organisational learning.	Next to the checklists and trade-off charts mentioned before, one could use decision matrixes and benchmark reports to visualise why a certain decision was made, which makes it easier to make a similar decision in the future.	Ch. 2 – formal quality systems Ch. 7 – risk identification based on previous experience Ch. 8 – standardised validation methods Ch. 12 – problem-solving

Glossary of Terms

Agile project management an approach developed originally in the software industry where planning and execution of projects is heavily devolved into self-organising cross-functional groups, in contrast to a micro-managed, centrally planned approach

ALARP as low as reasonably practicable

Angel investor a wealthy individual who provides capital, and sometimes expertise, to support an early-stage business

ASME American Society of Mechanical Engineers

Big Data data, typically customer usage data, which is so voluminous that traditional methods of analysis are not adequate

BoM Bill of material(s), a structured parts list that defines the composition of an engineering product or assembly

BSI British Standards Institution

Business model a company's plan for how it will generate revenues and profits, including what products or services it plans to manufacture and market, and how it plans to do so

CAD Computer-aided design, the creation of engineering information in digital, as opposed to paper, form

Capstone project an assignment that serves as a culminating academic and intellectual experience for students, typically at the end of an academic engineering programme

CEN European Committee for Standardization, Comité Européen de Normalisation

Collaborative R&D development work where two or more parties share their joint expertise in a planned programme of R&D

Co-location the principle of locating engineering, manufacturing, supplier development and marketing staff alongside each other for the duration of a project

Concept the basic principles of a new technology or product that can subsequently be developed into a fully detailed solution, usually arrived at after considering a number of alternatives

Managing Technology and Product Development Programmes: A Framework for Success,
First Edition. Peter Flinn.
© 2019 John Wiley & Sons Ltd. Published 2019 by John Wiley & Sons Ltd.

Consequence analysis modelling of the wider effects of a major failure, usually in a large-scale process plant

Cost of quality an estimate of the total cost of resources that are used to deal with internal and external failures and to prevent poor quality

Critical path the series of activities in a project that determine the overall timescale of that project

Crowd-funding a form of equity fundraising where companies seek a large number of small subscriptions to the equity of the company through an online platform

Design of experiments systematic experimental methods to determine the relationship between factors affecting a process and its output. It is used to find cause-and-effect relationships in order to understand a process better.

DFMA design for manufacture and assembly

Digital twin a digital model of a component, system, or process that is usually continuously updated with live data and is used to predict future performance and understand the operating environment more thoroughly

Duty cycle the operating pattern for a product when in service

Engineering change a modification to the formal information, e.g. drawing, used to define an engineering solution

Fault tree analysis a diagrammatic way of showing the potential causes of a top-level system fault

Five whys the principle of successively asking 'why?' five times when investigating the root cause of a problem

Flow the continuous processing of information and material without stopping, queuing, or backtracking

FMEA failure modes and effects analysis, a method of identifying and recording potential defects and their consequences

Fourth Industrial Revolution a new wave of major industrial development built around digital manufacturing, virtual modelling, embedded computing, artificial intelligence and autonomy

Gantt chart a time-based visualisation of the activities or tasks that make up a project

Goal-directed project management an approach to managing projects where the project is split into a series of intermediate goals that represent way-points on the route to completion

Hazard a potential source of harm

HAZOP hazard and operability reviews. Used mainly in the process industries, they are systematic searches for hazards that are defined as deviations to operating parameters that may have dangerous consequences.

Hoshin Kanri an interlocking cascade of objectives from the top of an organisation downwards

IMechE Institution of Mechanical Engineers (UK)

Innovation making changes in something established, especially by introducing new methods, ideas, or products

Intellectual property a wide range of intangible ideas or creations that might be expressed in the form of drawings, diagrams, or pictures and protected by, for example, patents or copyright

Internet of Things the interconnection via the Internet of computing devices embedded in everyday objects, enabling them to send and receive data

IPO initial public offering, an initial sale publicly of equity (ownership) in a business

ISO International Organization for Standardization

Kaizen continuous, small-scale improvement

Kanban technically, a card attached to containers of parts to regulate material flow but applied more generally to regulating the flow of information or material on a just-in-time basis

Lean thinking an approach to business management that emphasises value to the customer and that focuses on reducing waste in the processes that deliver customer value

Likelihood probability of occurrence

Loss function a concept developed by Taguchi to describe graphically how losses, e.g. customer dissatisfaction, increase as engineering parameters move away from their central value

Manufacturing readiness level (MRL) a measure of the state of preparedness of a manufacturing system for making a specific product, usually on a scale of 1–10

Mass customisation the concept of delivering personalised products to customers at close to mass manufacturing costs

Muda activity that consumes effort but creates no value; waste

OECD Organisation for Economic Cooperation and Development

OEM original equipment manufacturer. There is some ambiguity concerning the use of this term. It was used to describe the supplier of parts and equipment incorporated into a complex final product. It is now used more describe the company that integrates and assembles the final product.

Open innovation the deliberate use of inflows and outflows of knowledge to accelerate internal innovation and to expand the markets for external use of innovation

P2P peer-to-peer lending

Prevention the principle of avoiding the creation of defects rather than allowing them to happen and then correcting them

Private equity used generally to describe a situation where equity ownership of a business is not publicly quoted

Probabilistic risk assessment see QRA below

Process a series of actions or steps taken in order to achieve a particular end

Product development the development of a specific, commercial product for use in the marketplace, probably using a combination of existing and new technology

Product lifecycle management (PLM) systems integrated IT systems that manage all product-related data from concept design through to in-service operation

Product platform a common base of designs, systems, and components from which a range of different products can be derived

Project a piece of planned work or an activity that is finished over a period of time and intended to achieve a particular purpose

Project charter or mandate a one-page, agreed summary of the key points and objectives for a new project

Quality function deployment (QFD) a visual method of linking customer requirements to engineering features

Quantitative risk assessment (QRA) a numerical estimate of the probability of occurrence of major hazards

Research the development, for its own sake, of new ideas, knowledge, and science without any firm application in mind

R&D Research & development, a wide-ranging term covering basic research through to practical product development

Rework re-processing incorrect material or information

Right first time (RFT) the principle that it is always better to perform a manufacturing operation, or any other activity, correctly in the first place rather than doing it wrongly and having to correct or rework it

Risk the possibility, often expressed as a likelihood or probability, of loss, damage, or injury caused by a hazard

RTO Research and technology organisation, a public or private body carrying out technological research work and with the facilities and people to do so

SAE Society of Automotive Engineers, in the United States

Scalability the ease with which a design, facility, or business proposition can be adapted to higher business volumes

Science the systematic study of the structure and behaviour of the physical and natural world through observation and experiment

Scrums another agile project management term used to describe the team reviews that take place to plan or review sprints

S Curve a plot of cumulative cost or man-hours against time on a project, typically following the shape of the letter 'S', as the name implies

Seed funds a general term for early-stage equity funding by, for example, friends and family, angel investors, or crowd-funding investors

Servitisation development of an organisation's capabilities and processes to create better customer value through a shift from selling products to selling product/service solutions. Also known as 'Product as a Service'.

Simultaneous engineering engineering programmes where the technical work, market planning, manufacturing development, and supplier engagement are carried out in parallel with each other rather than in series. Also known as concurrent engineering.

SME's small and medium-sized enterprises, defined as having <250 employees and with limits also on sales turnover or balance sheet 'size'

Spillover effects in an engineering context, where an idea or technology developed for one application or sector is then adapted for an unrelated area

Sprints a term used in agile project management to describe short bursts of team activity to produce a defined, useful output

System integrator a company that engineers a complex product, such as an aircraft, by bringing together multiple products and systems from a range of suppliers, internal or external

Taguchi methods a suite of quality engineering methods developed by Dr Genichi Taguchi for use in the early stages of product development

Technology the practical application of research and science to develop new solutions that could subsequently be taken into commercial application through a product or service

Technology readiness level (TRL) a measurement of the maturity of a technology or product, usually on a scale of 1–9

Validation the process of confirming that a product performs as intended and will continue to do so reliably throughout its working life

Valley of Death the period, for a new technology or product, between initial funding of the idea and the point where sales revenue starts to be generated

Value proposition a summary of the benefits a product provides to a customer

Value stream mapping a visualisation of the flow of activities that create value in the eyes of a customer

VC venture capitalist, an investor who provides capital to facilitate the development of early-stage or growing companies

Virtual reality a computer-generated 3D simulation of the real physical world using typically a headset or glasses

Voice of the customer actual expressed desire for product functions and features

Waterfall model a V-shaped model that shows a product moving forward from concept to detail and then being validated from component through to complete product. The model is criticised as being unrealistically 'linear'.

Index

Managing Technology and Product Development Programmes: A Framework for Success,
First Edition. Peter Flinn.
© 2019 John Wiley & Sons Ltd. Published 2019 by John Wiley & Sons Ltd.